AF540298

# Industrialisation and Rural Development

# INDUSTRIALISATION AND RURAL DEVELOPMENT

O. P. Meena

*[The responsibility for the facts stated, conclusions reached and plagiarism, if any, in this book is entirely that of the Author. And the Publisher bears no responsibility for them, whatsoever.]*

**Industrialisation and Rural Development**



First Published : 2012

ISBN 978-81-8342-249-9

*Published by* :

**CRESCENT PUBLISHING CORPORATION**
4819/24, Mathur Lane,
Ansari Road, Darya Ganj,
New Delhi - 110002
Mob. : 9711991838
Fax : 91- 011- 23257835
e-mail : crescentbook@gmail.com

*Printed at :*
**Roshan Offset Printers**
Delhi

# Preface

The goal of any country is development. Development is development of people —their increased living standards and improved quality of life. Industry provides goods, services and material comforts to increase living standards. Social values such as human dignity, self-reliance and gainful employment for every person set the quality of life. Such development can only come by generating, mobilizing and optimally utilizing natural and human resources, natural genius and skills and maximizing the returns for the same.

Majority of the people live in rural areas. The question then arises as to how to maximize the returns, let us say, for a hectare of land. On land, we have soil, water, forest or agricultural crop, animals and people-under a given climate. How to maximize the returns for these resources is the issue. How to grow more crops, get better yields is one aspect. Soil-water management, better seeds, fertilizers, pesticides and farming techniques etc. have given good results. But these involved high-energy inputs. The emerging technologies like tissue culture, genetic engineering etc. offer great promise.

Past experience has clearly shown that rural industrialisation is not setting up large industries in rural areas. We are also fed up with the opiate that rural development is synonymous with agricultural development. True, rural people live on land and agricultural development is a must. But it is not enough. Agriculture should have a nexus with industry.

This book presents a cohesive and comprehensive scenario of the changing dimension of rural development process in the India, besides, producing a standard reference book for university students and research scholars. The problems of rural Industrialisation and development have been viewed in relation to agricultural and economic growth. The book suggests that there remains a lot to be done to brush up these attempts, because infrastructural disparities between rural and urban areas have accentuated the income disparities.

**Author**

# Contents

# 1

# Rural Industries

Village and cottage industries, except hand-looms, sericulture, handicrafts, coir, minor forest produce etc. have been treated as, more or less, the preserve of the Khadi and Village Industries Commission. Having given them the franchise, nobody else appears to have taken any significant steps to enable the commission to fulfil its role or complement its work by their own efforts to develop the sector. Except in Handlooms and to some extent in sericulture there has been no systematic work by the states or other organisations in this direction.

The Khadi and Village Industries Commission have been- suffering for long years by a running controversy on the limits of mechanisation. The arguments continue. Meanwhile the number of those seeking a living in this sector is obviously dwindling. In the running controversy over how much labour intensity is to be maintained and how much mechanisation is to be allowed, the artisan is being forced to continue the traditional labour intensive approach which involves drudgery sometimes of the entire family, with a small return in value added for the time employed. If we examine the Handloom sector or the leather sector we shall understand this basic problem.

The setting up of beams for weaving in Handlooms and the buffing and slicing of leather to prepare the material for production are two extreme cases of drudgery where long hours are spent and the value added for time employed is very small. In case of blacksmithy long hours are spent and many mandays lost in striking iron while it is red-hot with heavy hammers for forging. If these operations can be mechanised, and they can be, the artisan and his family can better employ their time in the actual production where value added for time spent is reasonable.

The phenomenon observed in the case of coir industry about the drudgery element and low returns is present in other pursuits to varying degrees. On the question of flight of artisans from their traditional pursuits one can venture a good guess that the processes involving drudgery without adequate returns, act as the basic retarding force. If this tendency is to be reversed, such processes will have to be suitably mechanised. Because of the historical background of the Khadi and Village Industries Commission almost having a monopoly in this field, this aspect has not received the attention it should have.

The Committee has no hesitation in recommending that the primary approach in the strategy should be to evolve a suitable mix of the manual and the mechanical for each of the traditional industries so that those parts of the operation which involve heavy drudgery and expenditure of time, without adequate value added for the time spent, are suitably mechanised.

Thereby the artisan and his family can use their time better in those parts of the process where their skill comes into operation and the return in value for time spent is reasonable and at least a living wage is assured. It is this intermediate technology, wherein skill is retained and upgraded, which has now to be developed consciously and quickly. The Committee recommends that this should be the key strategy for reversing the diminishing returns in employment from this sector.

The KVIC with its historical background has some justification for the conceptual difficulties in introduction of mechanisation, but the Committee wish to emphasise that those in favour of Village and Cottage Industries appear to be equally guilty of this conceptual confusion.

It is the fashion for any body supporting Village and Cottage Industry to add almost invariably as a rider that the approach should be labour intensive and mechanisation should be frowned upon. The Committee wish to point out that the justification for Village and Cottage Industries is not only the employment generated but the resultant improvement in quality of the consumer goods thereby produced.

This is the most significant factor which distinguishes the artisan production from the machine made uniformity and gives the greater justification for this sector to survive. The present position as explained in the key strategy approach is, that the artisan spends a lot of time in drudgery and repetitive operations which can well be performed by machines thereby making time for putting in the skill in the finishing of the goods and enhancing the quality aspects. If this can be achieved without putting people elf the industry, it will be the right strategy.

The Committee also emphasises that those talking of labour intensity to create an impression in the field workers that mechanisation of any sort is taboo. There has to be greater precision in what we mean. Can we reduce drudgery by proper mechanisation and at the same time not impinge on the employment opportunities in a significant way. Let us face the obvious fact that today, because of the drudgery and low return for the time spent, Village and Cottage Industries are fast losing their attraction for not only the present generation of artisans 'but preventing the new generation from seeking a livelihood. It is only by saving drudgery and fatigue and increasing the application of skill in production, producing better quality of goods, that the industries can survive and give a decent remuneration to the artisan. It is possible to save drudgery and provide necessary mechanisation leading to greater production, which as explained later, can be absorbed in the economy and thereby the artisan is not pushed away to other occupations. In case drugery and fatigue are suitably reduced and the production becomes available in larger quantities, the absorption of the products would depend on the viable growth impulses generated in the economy. The perspective would indicate hope on that front.

In this connection, the following extract from Shri B. Sivaraman's article 'Full Employment Mirage or reality' published in the December 1979 issue of 'Man and Development' is relevant in understanding the basic economics. As per the article, a ten per cent growth per year in industrial production is tenable according to present thinking. If every sector of industries keeps its relative share in the product, the informal sector can expect a minimum growth of ten per cent. In industries, we have reached "a stage at which we have to push our consumer goods, to effectively use the basic industries that we have developed.

The informal sector is purely one of the consumer industries. Further, the national policy seeks to give prominence to this sector hence it is logical to assume that we shall be working for a growth of much more than ten per cent in this sector. A 20 per cent growth is tenable. To absorb its relative share in the growth of the labour population two per cent can be reserved out of this growth.

To cover the relative share in the backing which is about 10 per cent, on the 1971 figures, a one per cent reservation should see the problem through in ten years. This reservation leaves anything from seven per cent and more for the production growth in the units. In ten years, a seven per cent compound growth rate doubles the production.

Thus we see that the strategy has enough elbow room for tripling or quadrupling of growth in this sector in twenty years. If the quality of the finished goods is improved and the necessary policy supports for the Village and Cottage Industries are introduced, there is no reason to fear that this sector cannot find its due share in the commodity markets. This supports our expectation of ten per cent annual growth.

In addition if the policy uniformly announced by all political parties that this sector will be enabled to get a larger share in the consumer market for the goods by suitable constraints on the mechanised organised sector is followed by the necessary controls, there is no need to question the possibility of this sector soon achieving a twenty per cent growth annually. If we achieve this, the arithmetic quoted above shows that the sector can maintain its employment potential and absorb its backlog of the unemployed and in addition absorb a steady 2 per cent and over from the increasing labour force (in the sector).

This leaves a growth rate for productivity of anything from 7 to 17 per cent per year in the sector. This means anything from quadrupling productivity per unit to increasing productivity by more than ten times by the end of this century. This can only be achieved by a suitable mix of mechanisation and manual skill. In fact a little thought will show that this intermediate technology of a suitable mix will have to be introduced immediately if we are to reverse the present tendency of deterioration in the sector. The approach is not only economically justifiable from the employment angle but also imperative for survival of the sector.

Within the frame work of the growing economy, the increasing demand for consumer goods would involve many new articles of consumption. This segment of the demand can safely be siphoned off to the sectors of production using intermediate technology.

These new lines of production opportunities will certainly attract the younger generation who may not like to pursue the traditional line but accept greater mechanisation which gives them better earnings. The data suggests, at different levels of manufacturing technology and for two points of time data for 1961 and 1971, a shift in the employment pattern from the low technological sector of house-hold industries to intermediate levels of organisation and production.

Non-traditional intermediate technology probably requires some mechanisation and a higher order of skills. The important inference, from the facts, is that the economy as it is absorbing more and more of labour in the intermediate industrial sector without any active push from the State. The experience of Japan is well worth a mention. Professor Ohkawa and Dr. Tajima well known authorities on the Japanese Economy, while dealing with the aspect of shift to new technologies in Japan had remarked that, among other things, Japan could move over fast to the new technologies requiring human skill because of the traditional arts of the village in many lines of production.

The Committee would recommend pursuing such a policy. The introduction of a new raw material, like plastics, sometimes introduces unhealthy competition from the highly mechanised sector to the detriment of decentralised production of household consumer goods, like leather, wood and metal works. In such situations, the aspect like cheapness of the artisan's ware against organised production is often ignored. The Committee would like to highlight the need to cover these serious deficiencies in providing the right impetus for development of this sector and effectively stopping contrary pulls.

Village and Cottage Industries do not require the costly cover of communciations, transport, housing, and water-supply and labour welfare. Energy requirements of course have to be met though a large part of the equipment is manual. Because the industry is plied in the house or nearby, mostly by a family group or local artisan group, these costly overheads are saved. On the other hand, the scattered nature of the production and the smallness of the individual units, creates other problems of conglomeration and consolidation for the market, except the extremely local. The needed support for the appropriate development should not be denied as without that the development may lag. The broad supports which can be identified straightway are:

(a) Continuous updating of the technology and moving towards higher productivity per unit by supporting research and development of intermediate technology which has low capital/ output and capital /labour ratios;

(b) Providing the training, design and market intelligence organisations so as to change the production lines from those for the purely local market to those can develop larger markets,

(c) Developing the necessary marketing organisations to collect and market the produce on a fair commission basis;

(d) Raw material supply for the industry at a fair price in small lots so that the enterprise need not have to invest large sums in inventories;

(e) Improvement of the tools of the trade so that the artisan is able to get the best tools that the latest technology can provide;

(f) Finding the necessary credit for the individual and the area to make the entire organisation work.

Many parts of the infrastructure are services which can normally be charged for at fair rates except in the initial period of development. Those in the nature of subsidies like supply of tools fall under the principle of capital transfer to the poor. In 1974 the Handloom. Committee, headed by Shri B. Sivaraman, suggested a state level covering organisation to give the necessary support for the supply of yarn at fair price and marketing the goods in the best market.

The subsequent development, resulting from the recommendation, gives ample proof of the validity of the strategy in finding an economic solution. The Committee, therefore, would like to focus attention on these aspects. It has already been pointed out that because of the wide franchise given to the KVIC, there is a complacency that no body else needs to do any thing systematically to handle and improve the sector. No doubt the Central Boards like hose for Handloom or Handicrafts and others try to fill up this gap to some extent, but they all suffer from the lack of an effective pervasive field organisation.

One approach would be to enable the KVTC to discharge its vast responsibilities effectively. The other would be to treat the KVIC as an available forum for the improvement of the sector but place the responsibility for overall achievement squarely on the State Governments who are the most directly concerned with rural employment and distribution of consumer goods. Studies ending with the detailed study of the KVIC by the Administrative Staff College, Hyderabad, shows that as of today there are inherent structural difficulties in the KVIC reaching the field level widely and effectively. This organisation has grown over more than twenty years and developed its own inertia.

In the judgment of the Committee any drastic changes in the working of this organisation will take a long time to be effective. As time is an important factor in the problem, the Committee is certain that beyond enabling the KVIC to play a more effective part, the main responsibility for rapid development of this sector has to be squarely placed with the State Governments. The role of the All India Boards can only be complementary as they lack field approach.

The type of organisation or organisations at the state level and the ramifications in the field and the complementary role of the KVIC and the All India Boards are being dealt with further in detail. The question of satisfying the preferences of the ultimate buyer of the products needs specific focus and attention through the development of designs, patterns, input supply, standardisation and marketing.

The case of handloom has already been referred to and in this instance the intention has not really been translated fully into effective action. Other relevant instances can be found in the case of match sticks where the State has not taken the relevant follow up action to cover the infrastructure of input supplies at fair rates, standardisation and marketing. A mere reservation is sold defeating unless followed by active support and cover. The example of Nutan Stove, in the marketing network, is worthy note. An all India consumption article of this relevant cannot be marketted by small units spread over the country without an active covering organisation.

We have failed in this and are trying to seek a solution in passing on this item of production to the organised large sector. The net result can be attributed to either lack in effectiveness of policies or the policies working at cross purposes and reducing effectiveness. An intermediate technology with a suitable mix of the manual and the mechanical is the requirement.

Though many people have been talking about the evolution of such an intermediate technology, as of today the only organisation which has done any systematic work in this field is the R & D organisation of the KVIC; but they suffer from, certain conceptual inhibitions. Even though better equipment is available in most trades, and better productivity is possible with existing technology, the level of equipment used and the output per unit shows conclusively how little the knowledge and the skill has percolated to the field.

Building up this vast organisation and infrastructure is a time taking job. There are a large number of industries and each has its own technology and skill. If we simultaneously seek to improve every sector, we shall flounder badly and will be able to cover none satisfactorily. The strategy of development has to be selective.

Out of the large number of industries, it should be possible to select those which at present engage a large number of the artisans. It is also necessary to select from present experience the new fields of growth where the production can be vastly improved and marketed. Sericulture, Tassar culture, Lac culture, Garment making including knitting of woolens, jewellery and stone—ware appear to be obvious fields with potential.

The fields should also be such that the intermediate technology has a prospect of developing quickly. The strategy should be to concentrate on the selected industries and during the next decade work out a time bound programme of action with the objective of increasing productivity per unit by five to ten times the present output. If a productivity level of five to ten times the present output per unit of the industries is to be achieved, within the next twenty years, with a gradually increasing growth, a mere mechanisation of the drudgery elements may not be the complete answer. Better equipment so that the skill may be more productive, and more widespread introduction of semimechanical aids, will have to be built into the production chain. Steps have to be taken to evolve urgently the structure for carrying out this intermediate technology on a crash basis.

There has to be a clear division of the research responsibility and a coordinating structure for bringing the scientists and technologists and the users together for development reporting, monitoring of performance and extension to the field for effective solutions. A rapid change of equipment, technology and training of the millions of artisans: in the field to enable them to utilise the technology requires a vast hierarchy of field level experts supported by a Pyramid of higher expertise and a large number of training units and organisations.

As of today, except for the technical experts in the KVIC and some field level experts in handlooms and partly sericulture, there is a tremendous lack of organisation and technical expertise in most of the village and cottage industries. As experience in the type of organisation and coverage is gathered if a constant watch is kept of technology development and new consumer demand in an expanding development and new consumer demand in an expending economy, other fields can be added to the list for detailed action.

A suitable forum for this continuous appraisal and change has to be postulated. Keeping in view the above principle, the committee suggests that in the Sixth Plan we may concentrate on the following industries: Food and tobacco products; edible and non-edible oils; beverages; textiles (khadi, cotton handlooms and manufacture of garments); leather, footwear and repair of footwear; major carpentary sectors; ferrous, non-ferrous metals; major items of production ' in non-matallic mineral products; sericulture and tassar culture; carpet making and woollen garments. Even within the selected industries, there is a need for selectivity in groups for development. The production units in village and cottage industries are generally family units with a sparse interspersal of Master Craftsman engaging labour from outside the family. These units are scattered all over the country in the village. The basic requirements of this family type unit is:

(a) Getting raw material in small quantities at a fair price in small lots through out the year so that the investment in inventories is minimal.

(b) Getting technical guidance in new technology and maintenance of this equipment at fair rates and promptly.

(c) Prompt marketing of his goods so that he Jin rotate his funds for raw materials purchase and. also meet his consumption needs by the value added.

(d) Getting the necessary credit at fair rates of interest for all these operations.

What is wanted is a covering organisation which can perform these functions for the individual family and replace the money lender/trader by a helpful and effective organisation. Such a covering organisation must also be economical, since a highly subsidised organisation will be inconsistent with the objective of an economic approach.

Taking into consideration the level and number of installments of raw material supply needed in the industry, the number of families a mobile unit can cover from a given centre and the minimum needs of a marketing organisation, each industry will have different norms of areas and number of families covered. Any attempt to go below this norm will lead to uneconomic costs. The strategy should therefore concentrate on a group approach for each of the selected industries.

The service unit will have to be located at a convenient centre within the area. At first sight it may appear as if this approach will leave out a large number of scattered units outside the frame. This is not so. A study was made of the spread of house-hold industries in some districts of Bihar. It was found that industries generally congregate in small groups of villages and large number of villages have no artisan at all or only a few. The Committee therefore recommends the group approach for each industry.

The basic strategy suggested by the Committee involves three crucial elements; first, upgradation of technologies to ensure using standards and quality production, second a covering organisation to provide the required support for raw material supply, marketing, credit and technology and third a group approach to ensure viability.

Besides this, the Committee would like to focus attention on the utter lack of a machinery to transform policies into effective action. The need for an effective thinking and acting centre analysing the parts for action and for further policy decisions is very desirable. This body should not involve itself in the execution but be a Brains Trust. Simultaneously, there has to be an effective structure to enforce the policies and give the cover for infrastructure needs and the training and marketing problems.

## Resources

The location and pattern of village and cottage industries have primarily been guided by the exploitation of raw material resources available in the rural areas and for catering to the basic needs of the people residing therein. Resources in rural areas are, by and large, varied and ample. The main sources of such resources are from agriculture, livestock, horticulture, forestry and minerals.

The sources of supply of other raw materials are many and are dispersed over a large geographical area.

By and large entrepreneurs and artisans manning the—village and cottage industries are unorganised and scattered over the whole of the country. In the absence of any worthwhile organised network, and weak bargaining power the artisans have to pay more for the purchase of the raw materials in competition to those from organised groups. Besides this, the weak financial base of the entrepreneur artisans adds another dimension to the problem.

On account of the low capital base the artisan is not in a position to take full advantage of buying his total requirement of raw materials when prices are low. If for any reasons the final product remains unsold for sometime, the artisan finds himself in a quandary and may have to resort to distress sale. All these factors combine to make the artisan dependent on money lenders and middlemen. On account of the vulnerability of entrepreneurs/ artisans, these people take full advantage of the situation and resort to the maximum amount of exploitation.

The dependence of artisans on money lenders or intermediaries for raw materials and marketing is linked with the operational modalities of the artisan. Generally, the artisan works on two broad patterns. In one of these the artisan works in his own home on piece wage on materials put to him by the middle-man. In the other pattern, the artisan is free to buy raw material and nay his workers wages from the advance given to him by the middleman. Under both patterns, the middleman purchases the finished products from the artisans and the artisan's dependence on the midleman is almost indispensable.

The institution of master craftsmen and their links with the individual family growers is also very much prevalent and this represents another form of useful though exploitative agency for the artisans.

In view of the predominance of middlemen in the raw material markets, the artisans find it almost impossible to purchase the raw materials straight from the growers. In many cases these growers are debtors of the middlemen and are obliged to sell their produce to them. Moreover, on account of being small growers, their staying power is naturally weak and they have to sell the produce on the farm itself to get ready cash,

The small şector loses most in its competition with the large in the price of raw materials for its production. The demand for raw material of an individual unit is small, even on the basis of an annual demand. Unless an industry can turn over its inventories three or. tour times in a year, the burden of interest charges will be high so the demand at one time will be still more small.

Perforce the sector has to deal with the retail dealer for its requirements and cannot take advantage of the low price seasonal purchases; whereas the large sector cannot only get its raw materials at wholesale prices, but being a large-consumer, can demand still more favourable rates from the trade and get it. As raw material cost as- percent of total cost, in the decentralised sector, is quite large, it is obvious that no amount of economic handling at the further stages of production can get over this intitial disadvantage vis-a-vis the large.

Attempts so far made to/get over this disadvantage appear to be few and far between. The main organiser in the village sector, the KV1C supplies cotton to the spinners and yarn to the weavers in the Khadi sector, but otherwise its supply of raw material to this sector of twenty-four industries is limited considering the magnitude of requirements. In handlooms, some amount of organisation has been attempted in forming cooperatives so that their demand can be combined and a competitive price obtained. States like Tamil Nadu have tried to organise departmental supplies to the weavers. Yet the cost of hand yarn to the weaver is nowhere near the price legitimately chargeable with the advantage in excise given to Hank yarn vis a vis the cone yarn to the organised sector and power-looms.

As at present, the State's intervention in the matter is negligible. Not only that, in the raw material supply to the carpenters and blacksmiths the State controls all timber at source and controls the supply of iron and steel scrap, and yet the artisan has to buy his material from the retail market at high prices. All the profits of the contractor and the trade are added on. The culmination of the absurdity is the case the Committee found in Muzaffarpur in Bihar.

The potter has often to buy his clay from the private farmer, even though the State owns most of the tanks in the area and clay got from desilting tanks is well suited for the potter. The trouble is that many of the State's tanks, it is reported, have been got recorded in the name of private individuals, or are under their control. Whatever be the covering organisation assigned with the functions, of buying and holding the stocks for dispersal, it is necessary to make suitable institutional arrangements for finding the needed funds on a priority basis at favourable rates of interest.

The large sector buys its inventories at the lower prices and holds on to stocks of goods to sell high in seasonal favourable markets. This requires a considerable amount of working capital which is advanced by the Banks in vv.-iys and means advances. Even the organised small sector docs not get this accommodation from Banks easily, let alone the informal and un-organised and village sectors. In case a covering supply organisation can be developed, regulated markets can be very helpful, for the village and cottage industries, in that the organisation can get the raw materials at low prices during the season, store them and release the same as per the demand of the artisans.

Here in raw materials does not imply only the basic raw 'materials hut also the one which arc outputs of one industry and become raw materials or others. In some cases such like production may be seasonal. The gathering organisation would have to take care of these products so that they can take advantage of buying cheap.

A similar sort of approach would be needed in the case of imported raw materials. The endeavor has to be to get the materials cheap to the production point. It may be worthwhile to have a glance into the nature of raw materials generally made use of in the village and cottage industries. Raw material cost constitutes a sizeable element of the total cost in the decentralised sector of the economy.

The First Five Year Plan while highlighting the important role to be assigned to the village industries, emphasised the appropriate utilisation of raw materials therein.. The relevant portion in the Plan reads- "As agriculture becomes more intensive, there will be greater demand for certain articles of consumption and tools and implements "which could be met by village industries,

The possibility of turning waste into wealth, for instance, production of gas from cowdung and other refuse of the village through gas plants in so far as the operations prove economical production of bone manure through bone digesters, soap making out of non-edible oils, etc., will further, provide scope for the development of village industries". In the Second Five Year Plan the idea of pilot projects was floated for the development of village and small industries and was advocated as one of the complementary activities under the communitv development programme.

One of the pre-requisites for selection of pilot projects was availability of local raw materials; the organisational pattern envisaged was co-operative in character and the mechanism of training-cam production centres was evolved. During the Third Plan period, based on the experience of the Pilot Projects, the programmes of Rural Industries Projects (RIPs) was introduced. One of the major underlying emphasis of this programme was the comprehension of intensive and integrated development of processing industries based on agriculture.

The RIPs were intended to promote the development of a cooperative agro-industry economy. Credit facilities were augmented for meeting the costs of raw materials, working capital etc. Development of agro-based industries as policy objective of the Fourth Plan was envisaged to have had a direct impact on rural industrialisation in the country. The National Cooperative Development Cooperation which had been set up in 1962, was given a fresh and vigorous base for strengthening the cooperative basis of agricultural processing, marketing and services.

Industrialisation of backward and rural areas attracted great attention and encouragement, as a means of achieving distributive growth under the Fifth plan and subsequent plans. Stress on production of mass consumption goods and optimal utilisation of local raw materials constitutes the backbone of development of village and cottage industries programme.

Besides the fragmentary approaches for providing suitable support system in the provision of raw materials for the village and cottage industries, a somewhat concrete system emerged in the working of the Khadi and Village Industries Commission. Since its inception the Commission had been giving due consideration in the provision of requisite raw materials for the artisans. The attempts have been rather on a modest scale

Over time the Commission had been able to evolve raw materials banks in respect of cotton, raw wool, edible and nonedible oil seeds and splints and veneers. The system created by KVIC needs to be developed and systematised so as to run it on commercially viable basis. The agency of District Industries Centres (DICs) envisaged in the Industrial Policy statement of December 1977, tries to comprehend a systematic approach towards the provision of raw materials for the village and cottage industries.

In the organisational set up of DICs. out of the seven managers, one was supposed to be specifically dealing with the raw materials. His functions included (i) to ascertain raw material requirenifnls of various units, their sources and prices: (ii) for cooperative or bulk purchase of raw materials by entrepreneurs, and (iii) to arrange for release of controlled raw materials. The loan facilities available under the erstwhile RIP and RAP Schemes had been proposed to be available through DICs.

The actual operational experience of the DIC's have been rather limited in character and as such any definitive conclusions are difficult to form at this stage, However, the limited field visits to two of the DICs in Bihar highlight that there are a number of operational problems and only a viable solution to these could bring about satisfactory operation of the system.

A review of the Working of Raw Material Depot (RMD) at Nawadha would be relevant in the present context. This grass root level experience in the operation of raw materials scheme for the rural artisans can help in evolving a viable scheme of action.

The RMD at Nawadha was in operation for the decade 1963-73. It provided assistance to the rural entrepreneurs in terms of (i) supply of indigenous scarce and controlled raw materials, (ii) supply of imported raw materials, (iii) supply of construction material and (iv) supply of improved handlooms and small machines needed.

As such the RMD had gone beyond its traditional task of supplying raw materials. The procurement of the articles was done in bulk and the artisans were charged only a nominal service charge over the purchase price. The activities of RMD enabled the entrepreneurs to purchase materials needed by them in small quantities as per their need and at cheaper rates, RMD was instrumental in introducing improved tools among the traditional artisans.

The RMD was provided' with a revolving fund of Rs. one lakh and during its initial operation for seven years, it not only met its establishment cost and sundry expenses but earned a profit of over Rs. 22000. The foregoing amply demonstrates that the policy and operational framework for provision of raw materials to the whole lot of artisans, dispersed over side areas, is not in tune with the requirements.

As at present hardly any regulation exists whereby the raw materials trade is obliged to make available, even the locally available raw materials, at reasonable prices. Price to the artisan is most important and it has to be viable. The Committee would like to stress this aspect at the outset. The wide dispersal of artisans and their weak financial position necessitates that their small requirements of raw materials need to be made available at the needed time and at their doorsteps.

The framework envisaged need to ensure that the financial, managerial and other covers are suitably provided for so as to achieve the desired goals optimally and effectively. At the grass root level, it is desirable for a start to provide raw materials to the artisan through the proposed Group Centre Approach. Dispersed artisans are very difficult to be approached in any meaningful fashion and hence concentrated groups of artisans are necessary for effective coverage.

However, any individual artisan located outside this group would not be denied any facilities if he so demands. All the same he would not get the facilities easily at his doorstep like those in the Group. He will have to come to the centre for getting the facilities. The Group Centre would help the proposed IDPA (Integrated Development Project Authority) in assessing the detailed re quirements of various raw materials and in also subsequently checking on the proper use of tin- nuv r materials.

The Group Centre would be the elective delivery point for the supply of raw materials to the individual artisans. The supply of raw materials to the Group Centres would be from the district level agencies. At the district level, the Committee recommends the formation of a DSMS (District Supply and Marketing Society) which would be given the responsibility for the procurement of raw materials. DSMS can buy when the going is cheap and supply the materials when necessary.

This society could ensure that the raw materials are stored at suitable places near the Group Centres so that the needed flow to the artisans is maintained. DSMS needs to be run on commercial viable basis on no profit and no loss. The inherent idea is not only to supply the raw materials but to supply the same at reasonable prices so that the earning capacity of the artisans is enhanced. The credit requirements of DSMS, can be met through commercial and cooperative banks.

State Governments 'can help b> providing appropriate margin money and treat the same as a development expenditure. DSMS should keep a proper rapport with DTC for optimal use of facilities. The Committee recommends the utilisation of the services of LAMPS by DSMS. Such like linkage would certainly be beneficial as the LAMPS art-operating at the grass root level and particularly so for the agricultural raw materials.

The activities of DSMS would be hampered without a suitable link up agency at the State level. Such an agency can help in procuring the raw materials which are not available within the district be that these are from other districts or imported. The Committee recommends that this state level agency be called SRIDC (State Rural Industries Development Corpo ration) and be responsible for handling all the raw-materials problems of the village and cottage industries. Village and Cottage industries and also the small scale industries suffer from the shortages of raw materials supply at reasonable prices.

As such the Committee recommends that a common organisation should handle the raw materials problems at the district and State levels. In case of any shortage occurring in overall supply of any of the raw materials, the normal tendency is that village and cottage industries become the first victim. A similar sort of treatment is there in respect of any imported item of raw materials needed. The Committee would recommend that requirements of village and cottage industries get the priority treatment and should become the first charge in that no cut be made in the amounts needed by these industries.

Such an assurance would go a long way in making the vulnerable artisans less prone to the vagaries of the open market mechanism operated by the middleman. The district and state level organisations would take some time to develop. Meanwhile in some industries like Khadi, Handloom etc. some system of raw material supply already exists.

The Committee would recommened that these systems should be co tinned and streamlined till the effective coming in being of the proposed organisations.

The Committee recommends that in case forest based raw materials, the forest department should be responsible for delivering the materials royalties plus transport to the DSMS from the nearest departmental depot, at the needed points.

The committee recommends the continuous monitoring of the process of availability of materials at the field level and its proper utilisation in production. Credit requirements for procuring and. supplying raw materials to the artisans will be fairly substantial. A rough estimate made by the Goya Org sat ion, the Committee has already referred to puts the demand between Rs. 30 and 50 lakhs per y Even though the ways and means credit can be obtained from the commercial banks or the cooperative system, margin money will be required for both DSMS and the SRIDC.

Sufficient margin money have to be provided by the State Government to the two organisations based on a reasonable appraisal the business that will be developed. Adjustment of margin money should also he made from time to as business develops. The Reserve Bank of India has been providing credit at cheap rates to the handloom sector which works through cooperative societies and cooperative marketing organisations. It is only reasonable to peel that similar cheap credit facilities will be given for raw material supply to all the other village in tries that are taken up for development by the States provided they are in the cooperative sector.

But problem is that the number of artisans that have be brought within the frame of development is so large and the Committee has already indicated that it is too much to expect formation of viable and efficient cooperatives to cover this vast number of artisans in any reasonable period. In fact, the Committee the view that expecting a cooperative structure could develop in any large measure in the near future will be unwise. The

DSMS and the SRIDC have been postulated by the Committee in order to meet this gap in organisation to this sector.

The cheapness of the credit is not really meant for the cooperative but is to enable the cooperative to give cheap credit to the artisan. The Committee recommends that the DSMS and the SRIDC who have to provide raw mat to the artisans at cheap rates should be enabled by the Reserve Bank to get its ways and means credit fc purpose at cheap rates equal to those given cooperative system. The Central Government already indicated that a Bill for formation of Na Bank for Agriculture and Rural Development introduced shortly in Parliament.

When such an organisation comes into being, it should be p for that organisation to support both the cooperative and the State Corporations on par for the necessary funds for supporting artisan classes. Till then, method will have to be found to give cheap credit the DSMS and SRTDC. The Committee recommends that this should be considered as a priority basis and an answer found early by the Central Government.

There are opposing views as to the process of technological change. One view is the Thomas A. Edison perspective. In this case, technological development is driven by profits. If a technology is profitable, it will be invented. The other view is that technology is a self-generating process. New technology is the result of old technology(ies) being recombined in new ways and used for new purposes. In the second view, profits cannot create the development of technology but determines its uses. What an individual perceives as a resource is influenced by the nature of technology.

In the 18th century, obsidian was an important resource among the inhabitants of the western United States; uranium was not. In the 21st century, obsidian is not normally regarded as a very important resource while uranium has become a resource. Factor endowment may influence the direction that technology develops. In a society with an abundance or arable land and a shortage of labour may produce (and consume) different goods and seek different technologies to produce them.

In the Edison view, the light bulb was invented because there was a demand for it and it could be developed and produced for a profit. In the second view, it is not possible to invent high-pressure steam engines, even though they may be profitable, until the technology of metallurgy develops metals to contain the higher pressure. Either view supports the argument that technology builds upon itself.

The creation of an internal combustion engine depended on its connections to cannons, oil, Maybach's spray carburetor, levers and gears. Each of these in turn depended on other technologies. When Daimler and Maybach built the automobile, it was the result of a series of connections between technologies that had been developed by many people over a long period. It is useful to think about technological change as a process. First, a piece of knowledge emerges or an "invention" occurs. Second, some one finds an application for the new knowledge (innovation) and uses it.

Third is the process of dissemination, i.e. the use of the idea is spread through out the social system. Each stage of technological change may produce or require significant changes in values and social institutions. Changes in social structure or the natural environment may encourage technological change. Technology and the social system are interconnected.

Technology has a strong influence on the structure of society and individual behaviour. The Industrial Revolution may be thought of as a fundamental change in technology of production that altered society. The development of the mechanical clock was driven by the clergy's desire to satisfy the institution of prayers at specific times of the day.

## India's Rural Markets

Rural contexts in India are essentially composite and digitally immature communication ecologies. Some researchers done a study on villages to analyse rapidly changing rural Indian socio-economic landscapes. Their findigs are briefly described in this chapter. For industry, India presents a good case study of communication technologies since there is much optimism in the country about their reception and integration into the everyday life of its citizens.

The social consequences of these technologies are often dependent on several causal factors and have partial and inconsistent impacts on populations. By some estimates, there are 150 rural PC-kiosk projects across India. Such projects could provide the first computing experience for as many as 700 million people in India. India's IT industry is seen as a singular route to opportunity and access to employment and livelihoods on the one hand and hip life styles and global cultural exchanges on the other.

Reflections of these structural changes cannot but be felt in small village communities either through infrastructural changes wrought on by the demands of a developing economy and/or the direct arrival of digital culture in their midst.

In India, where 'language, context, culture change in every few kilometers' potential IT users belong to a very large and highly diverse groups of which many are illiterate, a close look at these populations to understand the social context of software technologies becomes imperative.

Contexts that are receiving this technology need special attention to be able to see linkages between them and the reception patterns in its 'every day'. Recognising the complexity and diversity of Indian social landscapes were first steps to understand ICT applications in these contexts. Subject villages were one of the first recipients of ICT that arrived through a series of human interventions not exactly engineered by the recipient village or its communities.

To understand socio-cultural contexts of these recipient villages they undertook profiling their social demographics and communication ecologies. On village communication ecologies; It is evident that communication ecologies, especially in a developing region like India, are a composite mix of media, personal/ impersonal, formal/informal and has many people, media, activities and relationships interacting and evolving over a period of time.

It is important to map and record what is changing and partial in these ecologies. We were alert to the state and private initiatives differently impacting availability and access to media technologies. The spread of ICT's has added to the variety of media technologies available to people and debates around its impacts. Various actors have converged on the idea of communications technologies to augment development and business prospects for hitherto overlooked rural communities.

One of them is information and Internet access through a PC kiosk. Rural kiosks are computer kiosks in rural areas with one or more computers, generally owned and run by independent entrepreneurs. The researchers began by adopting an ethnographic approach to make sense of the complete range of social processes that come within the range of managing the business of PC kiosks.

Doing ethnography in these villages is to gain a perspective on the entire social setting and relationships seeking to contextualise these in wider social processes. The wider context, in this instance, can be eluded as a globalising economy affecting the villages in western India and offers up front the methodological challenges of ethnography to map rapidly changing social landscapes. Today's anthropological field in an interconnected world, when territorially fixed communities or stable local cultures are fast disappearing (the idea of isolated people living in separate worlds).

A field site, however bounded, is also an active recipient of larger political and economic dynamism and their fall outs. The state has a stake in ICT for development projects. Influenced by debates around bridging the digital divide in resource-stressed ecologies it launched several rural projects. These efforts were also underscored by the vision of bringing employment opportunities, infrastructural growth and community well being

The same debates around ICTs primarily driven by non-government organisations, focuses on ICTs for development and frequently point to shared access models as critical enablers of sustainable development and digital inclusion. In development discourses, communication technologies are viewed as development tools and technology becomes an agent of change leading to prosperity for a majority of citizens hitherto excluded from the fruits of progress. This vision of ICT's does not go unchallenged. Literatures have called attention to the challenges of national projects dedicated to digital equality for its citizens. Along these lines, a public-private collaborative effort has launched the ambitious 'Mission 2007-Every village a knowledge centre' for achieving a knowledge revolution in India

Technology innovators are major players in this arena. 'Disruptive technology', seemingly, was the key word in shifting the debate on low cost/high-utility technology for emerging markets and consumers. Disruptive innovation suggests that existing mainstream markets are not starting places for waves of growth, and there is need to "incubate technologies from ground up rather than introduce top down".

One particular initiative was a joint effort by engineering scientists in academia and industry. Faculty members at IIT Madras of the Telecommunication and Computer Networking (TeNet) group took upon themselves to pursue such R&D and found success and recognition. N-logue, a private company in league with TeNet, has introduced 'disruptive IT', setting up Internet kiosks in several rural parts of India.

Bringing ICT into virgin territories, for TeNet and N-logue, is not a government/NGO supported/subsidised process but linked to doing business with new groups, creating a business environment wherein the local unit can afford buying power and use technology profitably. For them, disruptive technologies will target the poor, drawing them within the market economy such that the transaction is enabling and empowering, and will create active agents in the circulation of capital, cash and material well-being.

The fact that rural India contributes significantly to the national GDP makes immense business sense to enable rural connectivity, while at the same time the Internet becomes an enabling technology. Private corporations took interest with equal enthusiasm, driven by both, the business prospects of ICTs in emerging markets and the vision of positive impact through their products.

The study has immense interest for Microsoft as it forays into emerging market economies making concerted efforts to engage with rural spaces. It is in the process of rolling out projects that take the benefits if IT to rural India to develop content and applications aimed specifically for the rural segment. The company is partnering with key players to accelerate the adoption of these services.

The interest in rural India is aligned with the overall vision of the Indian state and technocrats about their role in viewing IT as primary driver for social development. Microsoft India has decided to enter rural markets through a project called 'Saksham' (meaning self reliant) that will tap local entrepreneurs and help spread IT in areas that remain untouched by technology.

Saksham will help people set up information kiosks, tie-up finances for entrepreneurs through banks and involve local people for developing relevant applications for mass use. The company would initiate 50,000 rural kiosks in three years. Currently, six lakh Indian villages have around 14,000 kiosks, according to a Microsoft research .

Thus, powerful but often faceless institutions - the government, NGOs, technologists, corporations - all have a stake in the rural PC kiosk, which of course means that any study of the kiosk entrepreneur is incomplete without an understanding of the motives and policies of those institutions.

The subjects of field study were the 12 internet kiosks in rural India. A kiosk typically allowed customers, to browse, send mails, chat, offer on-line health consultancy, agri-consultancy, e-governance and on-line university admission.

Off-line activities include teaching basic computer courses, digital photos and web-astrology. E-governance services, often available in kiosks, issue relevant government documents and identity certificates to clients digitally, thus saving time, money and rendering the process transparent. The social origins of these kiosks were shaped by the following agencies; India's stated IT policies, corporates that were engaged in rural tele-com markets and evangelising, technocrats passionate about creating viable and affordable technologies for underserved populations and regions and developments pundits who were equally passionate about bridging the digital divide.

The 12 PC kiosks around Pabal village, Maharashtra, Western India, had a local person running each of its PC kiosks. The kiosks were linked to several players in the rural internet connectivity service providing tier. Vigyan Ashram, an NGO in the village of Pabal, operates as the of local service provider, LSP, for internet connectivity in village kiosks around Pabal, bringing ICT for the first time to these landscapes.

A private company, n-logue, in partnership with Pabal LSP, builds infrastructure for wireless internet connectivity. Both players are responsible for ensuring appropriate hard ware and soft ware packages, timely servicing and trouble shooting and a worked out financial arrangements by which all parties share and manage income from kiosks.

The partnership was the combined result of state initiatives bringing IT to rural regions, corporate vision to steer growth of kiosk/internet user markets, and the desire of local entrepreneurs to pull technologies and make home villages tech- savvy communities. Given the fact that technology arrived due to initiatives at a macro level, they found the business of running kiosks towards sustainability were largely driven by the KO's keen sense of business acumen and passion for computing technology.

They also sensed acute constraints in the form of the multi-party dependency in ICT ventures on extraneous players and agencies. The internet becomes a very expensive and frustrating experience to both owners and clients of kiosks when hardware break down and a minimum of nine hour power cuts a day anywhere in rural India being the norm.

The state, on the one hand, brings initiatives with much fanfare; make huge promises, dole out finance to kick start rural projects. On the other hand, they fail to persist to lend support or address the crucial goal of long term infrastructural assets for IT driven projects. All of these urged us to focus on the operators who were still holding on to their business and in some cases, finding creative ways of using kiosk resources.

Their initial work was around collecting base line information about the beginning of each kiosk, the motive behind investing in ICT, and the kind of financial and social support structures prompting this decision. They profiled KO's and social contexts in which they live and run their Kiosks. These include social positions of these individuals, family status, economic class/ landed status, educational levels and attitudes towards pulling technology into business. Apart from core interviews, they profiled the communication ecology and social demographics of villages hosting kiosks. This aided in locating the immediate and surrounding socio-economic contexts of kiosk business.

They recorded existing social structures of community life including communication. Details of village geography, social structures, economic/ agricultural patterns, water and electricity resources, migration, literacy/occupational levels and other demographic details were collected.

Recording village communication patterns and presence of mass media were crucial to the study to give us a sense of the demand for communication, news, entertainment and opportunities for kiosk business.

These included an actual count of telephones, land line and mobile, approximate readership of newspapers, cable TV connections, usage of postal services, estimates of audio-visual merchandise consumption and how do migrants keep in touch and transfer money.

Data was also collected to get an idea of popular TV channels, soap operas and mega serials. 12 villages, home to 12 PC kiosks became the situated anthropological field. Six kiosks make healthy business, four struggle but stay afloat, and two were temporarily shut down. Healthy kiosks have specific social geographies intersecting with industrial/urban belts that bring significant floating population with a need for internet services.

Two of these began by attaching kiosks to a flourishing business of teaching basic computer courses. One of these is a village bordering the outskirts of Pune city and four hours from Mumbai and part of a reserved green belt offering a unique opportunity for its residents to work in urban districts and live in a village. Most people in this village commute to work and it boasts of a railway station with 17 trains passing through in a day. The other village is again stationed on the fringes of an industrial belt, is a religious location drawing devotee tourists.

The third 'fringe' village with a busy highway splitting through is attached to a medical pharmacy. Around 4000 mobile phones, by rough estimates, are to be found amongst its population of 10,000-12,000. The fourth village of Kendur, needs special mention for its enterprising KO adapting to local demands, shifting kiosk services and tweaking technology to meet them.

The fifth village, in the healthy lot, transformed its socio-economic profile after agricultural lands were annexed by the government to develop an industrial belt. In the wake of new employment, in-migration and floating population, the kiosk began making revenues from internet services, e mail and chat. The KO, enthused by the socio-economic boom in his village, has plans to develop real estate, a shopping mall cum movie multiplex, to attract temporary inflow of potential consumers of this space. Elements of urbanity and porosity in village social contexts turn them to receptive agents of ICT's and create an organic demand for their services.

Steady inflow of new employees seeking new employment, encourage cash based market economy bringing in its wake the demand for personal communication devices and mobile phones. KO's in these villages manage to adapt kiosk business to new local demands for communication technologies. Incidentally, on of the successful KO is a dealer in mobile hand sets. Incidentally, all 12 villages have at least one computer housed in one of their schools, some having up to 10!

They are in various states of use or disuse as the case may be with some schools having trained teachers. Mapping communication ecologies of subject villages, made us notice inflections of urbanity and mainstream popular culture weaved into the social fabric of village communities. All 12 villages show very high TV viewership with significant private cable connections. A village with no cable TV service provider had 50 families sharing a privately owned dish antenna . Rural India's links with the mainstream (read urban) happens through social routes carrying popular cultural forms. Print, video and motion picture technologies aid and become prime mediating agencies.

The internet is understood from a popular matinee idol hero surfing the net even before the local school gets one. If particular village contexts proved successful providing a toe hold for ICT's, KO's in these villages initiate and respond to the intermeshing of new technology and village context. They are primarily situated in a socio-cultural world, be it an urbanising rural landscape or a mixed media environment with mobile telephony and satellite TV, affined to strong social networks as bearers and transmitters of information.

It is becoming clear from the village ethnography that technology and social contexts feed on each other to shape landscapes. Much depends on receptivity and social costs of technology to feed imaginations, prod human agency to learn skill sets, open businesses or take technology further to meet atypical demands. Operators find creative alternatives to keep kiosk business from sinking. KO's have shown immense enthusiasm in driving the PC kiosk business initiative and detecting commercial possibilities assumed non-existent.

These individuals are unique to their village environment in the sense that they possessed certain rare qualities. All of the KOs hail from farming communities with both or one parent practicing active farming. Most possess graduate/post-graduate degrees and have either studied or worked in urban centers. Out of the 12, 4 KO's are post-graduates, 6 graduates and the remaining 2 non-graduates having a basic diploma in computing skills. They are the first generation in the family to attain a college degree. They have made decisions, in some cases giving up active jobs, and reverted to native villages to start self owned business. They recognised in these individuals, pulling ICT into their village milieus, some specific and special personal qualities.

All of them are probably among the less then one percent of the village population who have a graduate degrees, worked in urban settings, have active socio-professional networks fetching them business and cheap infrastructural hardware and above all a tech-savvy sensibility that these personal/social experience provide to push them to adopt technology in underserved ecologies. Many of them articulated personal drive in bringing technology to their homesteads. The more enterprising the KO the more is the business coaxed out of latent demands creating active markets out of them.

They attribute persistence to a strong belief, glamour not with standing, in ICT's and their commercial visualisations. They managed business making most of what is available; car batteries, LPG cylinders are used to power the PC during frequent electricity cuts. It is interesting to see how desktop PC's attract other hardware/technology attachments to meet popular demands.

# 2

# E-marketing in Rural India

One of the biggest and growing economies of the world, India with the advent of Information, Communication and Technology (ICT) revolution, focuses its attention to connect its Villages to the world. At present, with nearly 40% of the Indian population being illiterates and 100 million school age children not getting schooling, there is a broad policy consensus that the only way India will get educated is if Information Technology (IT) is used to deliver it. It is the increasingly common these days to hear people saying that "The future is not what it was" because of the incredible advances and impacts of high technologies. We are undoubtedly on the road of a change in the way society functions and organisation itself.

IT is important because it represents parts of the fundamental wave of change, which is sweeping through societies. We are rapidly moving to information based economy and society. We are witnessing the growth of technology (IT), which has the capability to bridge the gap between the have & have-nots. India's capabilities in the field of IT provide excellent opportunities for the country to emerge as an IT super power in the world, only when IT is harnessed as catalysing agent for causing good governance and for launching a concerted Mosaic of action for causing all around socio -economic development especially in each rural sector.

There is great optimism over the potential for IT to support and promote Rural Development. IT can assist the rural population in areas like providing basic education, increasing healthcare services and accessing wellness development resources, providing training and skill development opportunities, increasing economic development and elimination of poverty by entrepreneurship/business community development.

Govt. and non govt. organisation are beginning to recognise the political and economic significance of the more than half of the world's population that lives in largely rural area and can be supported easily with lots of IT related tools. Rural development is gaining importance in India because of the majority of the population lives in villages. Development of villages is development of country.

The information technology forms the basic background for sound organisational system and public information utility. Communication on platform of existing network in the rural area has made in terms of its application and information sharing. It provides skills like group working, better communication, supervision and management of the available resources. The information technology (IT) is no substitute to any of the basic needs such as cleaning drinking water, basic health services, literacy, housing etc., however it has great inherent powers in speeding up the development process. It provides the technological advancement with social advancement in all sectors of the country.

## Rural Development Programmes

Rural population has been suffering from the un-availability of even basic requirements of human being. Be it a matter of drinking water, edible food stuff, basic health services, interconnection with road and rail network, education and many more, some villages in developing country like India even now are deprived of such utilities. The process of rural development is based on the agriculture mainly agriculture is heart of all rural development processes. Innovations due to latest technology are not available to the villages as there is no effective means of information sharing and utilisation in the real world.

This digital gap of communication between the knowledge base and the user can be eliminated by means of latest information highway. The reasonable accessibility and flexibility of the information technology (IT) may bring the innovative change in the mindset and the up-graduation of the major bulk living in the rural area.

Technology empowerment of poor is widely recognised as crucial element of any strategy of poor education. Information Technology has immense capacity to help the inequalities in society. The E-village concept is growing strongly with the help of government policies to enhance the development process of villages. The village may be designed around following some of the key features, which are associated with rural environment.

- Land and revenue records.
- Technology sharing with Internet and intranet.
- Training and refresher courses through web based learning.
- Contact program with virtual experts of various fields.
- Better governance and fast transaction of various functions.
- Networking and producer with remaining world-reducing the role of intermediate.

— On-line selling of village products like handicrafts, grains, vegetables etc.

— On-line market of various commodity items, virtual markets and review of appropriate market trends.

— Foreign trade transactions with latest information of the stock.

— Forecasting.

A realistic approach has to be worked out for model of resource sharing, collection development, networking etc with due consideration to the present communication and computing infrastructure. The realised representation is shown in the figure.

The model is based around the establishment of Datacenter at District and Block level in phase one that will be connected with state & National Data center. The basic function is to form an effective line of communication network between the accessibility of the external services and the solution of the specific problem associated with the rural environment. This can be achieved by means of information database.

A very effective review process for the verification of the output results is essential for the optimal performance. The collective database in the form of feedback must be interacted with information available to make suitable changes or modification for the appropriate action.

## Rural Health Care Services

Many public health problems can be prevented or treated through information dissemination (e.g., through remote diagnostics), often at lower cost than treating the problem afterwards. One of the many challenges facing the health care sector today is the adequate and affordable provision of specialised health care services in rural or remote areas. For example, radiology (X-RAY) services. This challenge can be met by using Information and Communication Technologies and tools such as a Teleradiology system to explore how advanced technology could support the delivery of health care to isolated rural communities.

One of the major problems which members of the public and rural masses face is to get access to the land records. In many countries, this has been one of the priority areas for application of IT. The decentralised databases can hold the local level land records allowing easy access through dedicated terminals. The transfer and registration of deeds can also be made much easier. Therefore IT revolution is reaching to the doorsteps of ordinary citizens. This will also drastically reduce the number of litigations arising from land-related disputes.

Research shows that there is a wide spread lack of market coordination in the rural areas of developing countries. Information can play an important role in enabling market coordination. The information and integration can help the poor by promoting elimination of poverty through economic growth.The vast majority of the poor in rural areas of low income countries are either farmer or surplus laborers. Each group depends on markets, and thus can be assisted through better functioning markets supported with Information Technology.

The farmers primary interest is to maximise the profits they earn from theirs farms. To achieve this goal, farmers need information for the important purposes given below

- Relative prices-allow the farmers to make decision on the mixture of crop, its price tells him how much to produce.

– Price information enables them to produce in a more efficient manner. They will be able to purchase inputs (e.g., fertiliser, irrigation equipment) at cheapest rate. Prices may also alert them the existence of inputs that would profitably boost their production. Price information also allows them to know where to sell their output.
– Proper Information provides them the suitable type of cultivation and associated profitability.

## Rural Entrepreneurs

Rural entrepreneurs may face greater hurdles in accessing capital that those urban, because of the perceived novelty of their request and because rural entrepreneurs may be unfamiliar with formal financial institutions. Nonprofits, such as India's Dhan Foundation, have developed programs to assist their self-help groups in applying for loans from banks and government programs. Mechanisms and legal provisions for billing, settling accounts, issuing credit/smart cards, and transferring funds determine the appropriateness, cost, and quality of certain services (e.g.-commerce, national and international remittances). This business backdrop is a combination of government policy, the legal and regulatory environment, and practices within financial institutions, and therefore depends on diverse stakeholders to ensure its effectiveness.

Rural children miss primary education. Many educational experts are convinced that the only way quality education can be delivered to masses, cost - effectively, is through information technology. The authors advocate developing a program to train villagers core group (VCG) including women and enthusiastic youth, which will ensure that IT reaches the common person even in remotest part of the country.

Proper staffing at various levels is an important requirement for successful implementation of the program. At state level, one Project Director can be appointed who will solely in charge of the Information network.

A project officer can be appointed at the district level, to monitor the progress Of program and to send periodical report to the state government and down to village level include Block Information Officer to control and regularly monitor this program. The government should set up an information kiosk in each village consisting of a Multimedia PC, Printer& Internet connection with local language software. When the Village Core Group (VCG) completes their training, they can take charge of the information Kiosk of the village / neighboring villages.

The service provided by such Kiosk is multifarious. The information about market rate of products, agriculture knowledge, e marketing can be available to farmers without any external intervention. This will definitely help farmers in getting a fair price for their products. The generic information found in the network should be rendered into locality specific knowledge on which farmers families and rural women can act. The basic job of the information kiosk can be follows:

- To enable rural families access ICT.
- Training rural women and rural youth in operating Information kiosks.
- Transform generic information into locally relevant information.
- Maintenance updating and dissemination of information on entitlements to rural families using ICT techniques.
- Building of a model in information dissemination and exchange among various VCGs in the District.

- To provide information on government schemes offering employment or on educational opportunities for the children.
- Establishment of the database consisting of reliable information to keep them fully informed of what alternatives are available to them in terms of hardware, software and technology.
- Development of the integrated set of websites and content management tools using multimedia tools available in the local languages.
- Development of workshop and variety of training program to strengthen local capacity for use of IT as tool of empowerment.

Another suggestion is that various organisation and NGOs and Government agencies should allocate funds to engineering colleges to conduct rural uplift programs. Staff and student chapters can provide computer training to village school teachers. The student of engineering colleges can do their vocational training at villages and should prepare project report about upliftment of rural areas through Technologies. If each technical college in the state adopts a village and imparts such a training strategy, it will definitely bring the Indian villages closer to the rest of the developed world.

## ICT Provision in Rural Areas

Given such trade-offs, there is a need to identify which kinds of ICT access deliver the best value for money, and how the limited resources that can be spent on it can be made to best suit the particular needs of rural India. A number of 'models' have so far been tried. One of the most famous projects of successful ICT application for development is the Grameen Village Phone system, undertaken by Grameen Telecom (a member of the Grameen Group). The project aims at ultimately spreading phone access to the over 100 million inhabitants of Bangladesh who are so far 'unwired', made possible by combining the Grameen Bank's expertise in village-based micro-enterprise and micro-credit with the latest digital wireless technology.

The aim is to have selected member borrowers of Grameen Bank purchase the phones under a lease programme and make the phones available to all users in the village on a fee-paying basis. Another model of ICT provision in rural areas of developing countries, and one which attempts to combine phone access with access to the Internet is that of the so-called Telecentres or Information Kiosks or the recently introduced Infothela of MLAKLH.

An Infothela is a common point of access for multiple users (often an entire community), providing a range of ICT services including Internet, fax, phone, e-mail, word processing, and even specialised information retrieval or applications (e.g. distance education or matrimonial matchmaking). Telecentres have been established widely in the developing world, and vary in their service provision and means of funding. In Peru, the establishment of numerous 'Cabañas Públicas' created one of the highest concentrations of public internet access and a significant reduction in prices.

Nevertheless, the experience with telecentres has so far been a mixed one. In numerous cases, usage, particularly of PCs, has been lower than expected or commercial viability was not attained. Of the over 70 Community Telecentres established since 1997 by the South African Universal Services Agency, only 40 per cent remain open today, with only 3 per cent making enough money to cover costs. Buried at the end of the World Bank policy paper on the 'Networking revolution: opportunities and challenges for developing countries' is an account of multipurpose community telecentres

(MCTs) in rural Mexico. It turns out that of twenty-three MCTs built in rural Mexico, only five were working two years later. This is a failure rate of 80 percent.

The policy paper comments, "Problems encountered included insufficient maintenance funding, inadequate political interest and will, and cultural constraints which hamper community interest in the projects." The paper gives no hint why "political interest and will" might have been inadequate and why community interest might have been constrained by that holdall excuse for failure, "culture."

The paper concludes that the Mexican case "underscores the importance of participatory design and attention to sustainability issues in the development of such programmes." Internet and Information Kiosks exist in various kinds, each with their respective merits. First, one might distinguish between the small private sector cyber cafes on the one hand and bigger, donor-funded telecentres like e-Seva in Andhra Pradesh or e-Village in Pondicherry on the other hand.

Smaller, privately run cyber cafes are often financially self-sustaining—but are thus usually restricted to areas where they expect to be viable (usually urban centers) and are usually neither within physical nor financial reach of the poor. They are also unlikely to be able to provide local content.—By contrast, larger, often externally funded telecentres are rarely financially sustainable but can focus more on specific 'development'—aspects, including access specifically targeted at rural communities and the poorest in general, as well as a focus on training.

A second distinction is according to the institutional context they are embedded in. This often has a significant influence on the 'developmental impact' of telecentres. Commercial telecentres and commercial franchises (like e-choupals of ITC) are usually closest to commercial viability but, as mentioned, are unlikely to have an impact on the poor outside the economic circle of the e-Choupals.

Telecentres run by or with the involvement of developmental NGOs are more likely to target poor and marginalised communities and focus on much-needed additional services like training, content creation, provision of public goods without which ICT access would be of limited developmental use. Telecentres in village schools for example as another alternative have the significant advantage that for their establishment an existing physical infrastructure only has to be extended and some of the ICT-relevant training can be cost-effectively integrated into the mainstream curriculum of the educational institution. This partnership has successfully worked in the DM project.

One further idea for the Digital Mandi that evolved was Virtual Telephones or village voice mail systems, as have been set up in Brazil. These can provide individuals with their own telephone number and access to a voice mailbox. In other words, the individual need not possess a telephone but can receive calls to a voice mailbox using his/her personal PIN.

Extending this idea to text e-mail access, a South African company assigns e-mail addresses to every Post Office box address in the country, thereby providing electronic mail indirectly to around eight million South African households through public internet terminals located in post offices which users can access with a personal identification number. The Postal Department in India has now taken up a similar programme.

Thus there are a number of alternatives and apparently mutually exclusive business models for ICT implementation in Rural India. On one hand it appears that kiosks run by local entrepreneurs with localised and targeted applications (like e choupals in Northern India or voicemail service in Brazil or matrimonial matchmaking service in Tamil Nadu) will succeed on the other hand following the success model of the world wide web itself one may suggest that if an infrastructure is created and user friendly appropriate interfaces are continuously accessible then local rural folks will develop their own applications and Information Kiosks or Infothela will survive.

The Digital Mandi Project conceived as an electronic trading platform for agro-commodities of Northern India and meant to run as a core application on the mobile Infothela faced additional problems. Barriers to information access may be physical, economic, intellectual or technological, that impede a user's participation in the activities on a website.

The barriers may be actively imposed by the architects and website designers or they may be allowed to continue simply through their lack of action or lack of understanding of the critical user conditions. Such critical user conditions may arise due to particular demographic, geographic, cultural, social, psychological, economic or other factors.

Issues related to usability such as ease of use, usefulness, decision effectiveness, user response, user satisfaction and many other aspect of usability have been studied in great detail by researchers. But interactions with focus groups at various agricultural market places around Lucknow-Kanpur showed the need of a more detailed study on Information communication barriers on a more localised set of priorities.

A general framework for web design keeping in mind the human-computer interaction theories, web site usability principles, information intensity paradigm, e-customisation models and heuristic evaluation models is already in place and is assumed to sufficiently address the question of defining broad guidelines for designing any successful website.

It is therefore, assumed that a website with relatively high-level of accurate, up-to-date and pertinent content, deployed in a user-friendly way, customised to particular user groups, and tailored to specific geographical needs should be universally successful and hence, accepted in India too. However many such efforts have apparently failed to achieve their targets. The challenges to agricultural website usability for rural marketing in India arise mainly because of the highly specific local needs and the great diversity in local conditions. The major challenges are

- Poor literacy rate—low use of textual information
- Remote village locations—physical distances compounding problems of lack of proper price information and habitual dependence on middlemen.
- Absence of alternate media for dissemination of info.
- Absence of info in vernacular languages and multiplicity of languages.
- Cash crunch of farmers, immediate cash transaction system and reluctance of banks to provide soft loans to farmers.
- Economic, low-cost solutions—any technology solution aimed at benefiting the masses in rural India must be affordable and low-cost so that the perceived economic benefits of such an endeavor are much more than the cost of switching over to a different technological solution.

In the absence of timely and correct information about prices, arrivals and market trends, compounded with the problems of low cash-at-hand and proper advice, farmers are forced to sell their produce at lower-than-expected rates. The result is that the benefits of the 'green revolution' have not really percolated down to the farmers. The Digital Mandi project demonstrates that there are a number of features pertaining to the ICT access projects that are particularly successful from a 'developmental' viewpoint This means, for instance, to somehow convey the relevant (local) content provided through internet access to the largely illiterate rural populations of developing countries in local language may have far- reaching spin-offs.

A model that inspired the Mandi team was The Kothmale Community Radio in Sri Lanka. This project has combined community radio and Internet access. It has a leased line connection to the Internet, and in the so-called process of 'radio browsing' programme presenters browse the Web in the studio on behalf of listeners (who provide requests/input through phone or post). Relevant 'experts' from the community then interpret the information for listeners.

Another good example of the creation of relevant local content are the 'Infoshops' in Pondicherry, India. After information requirements are identified during a trial period, volunteers from the village create a local database comprising government programmes for low income rural families; cost and availability of farming inputs such as seeds and fertilisers, grain prices in different local markets; a directory of insurance plans for crops and families; pest managements plans for rice and sugar cane; a directory of local hospitals, medical practitioners and their specialties; a regional timetable for buses and trains; a directory of local veterinarians, cattle and animal husbandry programmes.

All these preceding experiments contributed to the Digital Mandi design. Web site success depends on a number of factors the most important of which is the 'website design', which encompasses both the content creation and information design.

All website design issues aim at providing certain requisite features in the website. Jonathan Palmer has contented that website success depends on such factors as website download delays, navigation, content, interactivity and responsiveness. The inclusion of these features into a good website can be addressed as (a) content-related issues, (b) information design issues and (c)communication design issues. Traditionally, the basic work of a good website design has been considered as addressing the content-creation and context-appropriateness issues only.

While website designing objectives for other purposes may be fulfilled by creating a good fusion of relatively high-level content with fine design features taking care of the issues mentioned above, they would certainly fall behind their objectives when considering agricultural websites for rural marketing. The reasons as has been listed earlier, can be found in the inherent characteristics of rural markets in India. These websites are therefore bound to fail unless delivery services for agricultural information can be effectively integrated with good information design models and grass root innovators/social activists' agenda.

The digital divide, which we can broadly define as the unequal distribution of the benefits of the rapid technological advances of the last decade not only among rich and poor nations but also across social groups within a given society, is experiencing one of the most spectacular reversals in rural India. That rural India, with 70% of the country's population of one billion, would become one of the hotspots in this transformation seems far-fetched given the country's lag in basic infrastructure and low income. There are only 3 telephones per 1000 rural inhabitants, 65% of the population is illiterate, and the average income is only U$1 per day.

Despite this, there is a market for digital services in villages. For example, to sell their produce, farmers traditionally rely on traders who are known to quote rates far below the market price and pocket the difference. Farmers otherwise have to travel long distances to find a wholesale market that offers more competitive rates. It is also not unusual for villagers to travel to faraway government offices at district headquarters to submit applications, meet officials, obtain copies of public records, or seek information regarding prevailing prices in commodity markets.

This involves the loss of a day's income as well as the cost of transportation. Once at the government office, the relevant official, record, or information could be unavailable, forcing repeated visits and additional expense. Villagers also face discomfort, harassment, and corruption from public officials, or they are simply not given or do not understand the correct market prices. The government clerk or officer-in-charge works from paper records and has a monopoly over the information and records, leading to inefficiency and a lack of transparency.

Wiring up the farmers' villages may seem like an unusual way to help them with their problems. Nevertheless, studies show that there is potential for digital services rural settings. It is a matter of providing the right kind of services with reasonable cost, and overcoming poor power and communications infrastructure.

## Rural Indian Digital Market

A number of projects have been launched, with the sponsorship of both public and private entities, offering a limited breadth of services such as information kiosks, e-governance and internet access. Companies/projects include: Gyandoot, Rural information centres in Kerala (an Indian State), Info kiosks, Warananagar project, SARI, TARAhaat, Drishtee, and EID Parry.

A number of different business models are being tested all over the country with the common objective of bringing IT to the masses and empowering the world's poor in the process while creating a viable business proposition. Through cyber kiosks various services are being provided to villagers. Primarily these services include E-Governance, entertainments, e-horoscope, computer education, health and agriculture related information, market rates, weather information etc.

Following host of issues are seen as the common challenges faced by these projects:

- Partnership model - understanding the potential of partnerships with third parties
- Business model - right mix of services, revenue projections, aligned incentives
- Technology - how to deal with the lack of power and telecom infrastructure, and poor technology penetration
- Efficient back-end - operations and buy-in from content providers. However Drishtee, through its profitable kiosk business, has proven that significant demand does exist.

Drishtee is a socially oriented company seeking to bridge the cultural divide in rural India by delivering on- and off-line kiosk-based services. Drishtee plans to build an intranet platform that would be scaled to offer a wide range of information, services and goods across all of rural India.

At the current stage of the project rollout, Phase I, the firm is facilitating a limited offering of government services. These services include basic government transactions, for example, obtaining a driver's license, filing complaints relating to government-run services, and obtaining below poverty line, marriage, and death certificates. In addition, Drishtee offers computer training in its kiosks. The rates charged by Drishtee average 20 rupees (40 US cents) per transaction and 300 rupees (U$6) a month for computer training.

In Phase I we have identified three factors that contribute to the success of digital divide solutions and the viability of the business model:

- Create an economically sustainable business model by addressing local needs and utilising the appropriate technology within a specific geographic, economic and social context.
- Extend value proposition to a large customer base in order to realise efficiencies and economies of scale and scope within a broader geographic, economic and social context.
- Replicate and franchise core sustainable and scalable aspects to similar areas.

*Network Management Company (Drishtee)*: This is the role performed by Drishtee, at this level the network is planned and managed. The roles of the network management company are to create the network by demonstrating the business success of the district operators and recruiting partners and to provide ongoing support by managing and updating software and interfacing with government entities and corporate clients.

*District Operator (hub)*: The business plan foresees the recruitment of 550 district operators (1 per district). The responsibilities of these operators include building the network at the village level, which requires recruiting and incubating local entrepreneurs in each village as well as providing backend support to the village kiosks and interfacing with the local governments.

*Village Operator (kiosks)*: A total of 55,000 village operators is targeted throughout rural India (100 per district) and its key role would be to market and deliver services to villagers.

The network would be supported by two sources of revenue: a one-time licensing fee and a stream of transaction-based fees. The village kiosk would pay the one-time fee to be split between the district and the network operators and share its transaction-based revenue with the other tiers. District hubs also pay a licensing fee to Drishtee.

Based on aforesaid model, Drishtee plans to earn money through one time licensing fee and a percentage of transaction fee for various services. The transaction fee ensures continuous inflow of revenue to the company. The deficiency in communications and other basic infrastructure in rural India creates an opportunity to establish partnerships with corporations active in the agricultural Indian industry. From trading companies to capital equipment providers, these firms would greatly benefit from the creation of a network that would make the Drishtee business model economically sustainable.

## Handheld Model Services

Handheld devices can be used to provide various personal and group related services. It is important to note that the entrepreneur (the "PDADoot") visits about 10 villages surrounding the kiosk on regular basis. Following is the list of paid services that handheld model can provide.

*Personal Paid Services*

- Camera and Printer combination.
  - Picture taking and printing.
- PDA without internet connection.
  - Picture viewing once taken with digital camera.
  - Inter-village communication (voice mail/email).

  - Performing various tests with appropriate equipment, ie: health related and soil tests, etc.
- PDA with internet connection
  - Email.
  - Voice mail.
  - Internet browsing (cached from kiosk).
- Scanner/Camera and Printer combination.
  - Scanning and printing formal documents.

Apart from personal services, PDADoot can also deliver group services for the whole village. For group services, the panchayat pays for the services.

*Group paid services*

- Weather Information.
- Mandi (Market rate).
- Government Schemes.
- Post local jobs and ads etc.

In order to generate awareness among villagers, it is important to provide free services. In addition, free service will help the entrepreneur to be able to build up relationships with its customers (villagers), and thereby increasing its customer base. Following is a brief list of personal and group services that can be delivered free.

*Personal free services*

- First five emails/voice emails are free.
- Take the picture and see it on television /PDA, etc.

*Group free services*

- Community Health Awareness
- Community awareness about education/ employment news, knowledge about population control

— Transportation schedule
— Agricultural information

**Handheld Model**

The entrepreneur who runs village kiosk should own the handheld device. Depending upon the demand, he/she (or his/her family member) visits the surrounding village at regular frequency.

On the entrepreneur's visit to various villages, he not only delivers various services but also collects requests for specific services for which he needs to get information from Internet. After coming back from villages, he downloads information from web on the PDA (while also taking printouts of various forms) and in the next visit he delivers those services.

Since the PDADoot moves from village to village, it's important for PDADoot to be a male member of the community (for most parts of the country). Apart from being literate, a PDADoot must be chosen who does not discriminate nor is discriminated by people of different caste.

Mobile services do add value by:

— Generating greater awareness of existing services.
— Penetrating more broadly within and surrounding the village geographic market.

The PDA helps reduce the complexity when providing some of the services, like email/voice mail, E-horoscope, Government online etc. At the same time, by providing door-to-door service the demand for these services also rises. In addition, handheld devices can also be used to provide following three services:

— *Inter-village email/voice mail:* The PDADoot can record email/voice mail on the way from one village and deliver it to the next village. It is pretty common that these villagers have relatives/friends in the neighbouring village.

— *Information Collection and Marketing:* From time to time the government needs data collection from villages and invests a lot of money. With the Kiosk model and Handheld PC (PDA) service, the PDADoot will collect the desired data while going from one village to another. In addition, if government wants to give information to villagers it can be distributed to the kiosks. These kiosks coupled with PDADoot could also help companies such as tractor or fertiliser manufacturers to market their products among villages.

— *Photograph Related Services:* Handhelds coupled with inexpensive digital cameras ($80~ Rs 4000) can provide affordable photography services. Traditionally, these villagers go to camera shop in the near by city, get their picture taken (which they can't see before they can be developed), and then again revisit the camera shop to get their pictures after some days. The whole process costs them money in transportation fare and money charged by the camera shop person. In addition it costs their time, by losing several days of work. In the handheld service model, the PDADoot will use a digital camera to take their picture, and then he can show the picture instantly with the help of PDA, and get the approval for printing from the client (villagers). In this way they can also see if a satisfactory picture has been taken. The whole process takes relatively less time and money.

The team also learned that whatever be the theoretical findings, it must be tested in the field. The reason being rural digital market has not been tested and hence the final effect can only be known through sample field studies.

**Investment in Handheld Services**

Handheld devices offer the opportunity to extend kiosk-based services as well as handheld-specific services to a broader customer base both within and surrounding the village - thus expanding the market reach of the kiosk. In order to assess the value of these handheld services, we constructed a model identifying the success criteria for generating a positive net present value within 5 years. The model indicates that handhelds make a successful addition to existing kiosks if they can generate transactions from 7-8 new households per week, a figure which appears to be realistic:

- No additional labour costs would be required to transport the handheld around the village and deliver the service. This is based on assumption that village kiosks are run by families with under utilised labour capacity.
- Handhelds can generate a total of $0.65 per household from new services such as photos and voice mail.
- Handhelds can replicate 50% of existing kiosk-based services and corresponding revenue - $0.30 per household.
- The handheld operator will to travel around the village to deliver the services for $10/week.
- 10% hurdle rate for the initial capital investment required.

As prices of handheld devices continue to decrease according to Moore's law, the one-time and ongoing costs will decrease, and the model will only become more attractive. Based on these assumptions and data points, we recommend an investment in one handheld for each kiosk to penetrate more deeply into the rural market environment.

## Rural Service Provider in India

The lack of 'access' has curtailed their ability to compete. In fact, ingenuity and hard work has not been adequate for one to enjoy economic and social benefit. In order to acquire these benefits, access to resources like education, health and employment become critical. The Internet has been a boon in this regard. Today, one can be in the remotest corner of the world and as long as there is access to Internet enjoy access to education, health and resources. This allows them to compete and use their ingenuity and hard work to bring about a significant difference to their lives. The total rural population of the developing world is about 3.5 billion, with their average per capita income being no more than $ 200 per year. India, with 700 million rural people located in 600,000 villages, is a reflection of the developing world.

India has over 700 million people living in over 600,000 villages. Therefore any programme that is implemented in 100, 1,000 or 10,000 villages is miniscule and makes very little difference to the entire rural economy of India. The underlying element here is that a programme should have the potential to scale to half a million plus villages. Anything that is not scalable is simply an experiment, probably worth watching, but not of significant relevance.

There are three factors that are mandatory to build a scalable and successful business in rural areas.

These are:

- Technology that is cost effective, affordable, robust, scalable and capable of delivering the relevant applications
- A clear business model that addresses all market, stakeholder and operational needs
- An organisation that is exclusively focussed on the rural market, which thinks and acts rural

**Connectivity**

Over the last 15 years, the state owned incumbent telecom operator, BSNL, has taken fibre to almost every taluka (county town) in India. These fibre lines are infinite bandwidth pipes. Further, if a wireless coverage of 15 to 20 km is established from these county towns, one can cover almost all villages of India.

In India, a taluka typically has 300-500 villages and most of them fall within a 30 km radius. What is even better is that wireless technology is constantly evolving - costs are coming down and bit rates are going up. In other words, technology is present to carry this task forward. The corDECT technology, jointly developed by TeNeT (Telecommunications and Computer Networks Group of IIT Madras) and Midas Communication Technologies Pvt. Ltd, has proved to be a major breakthrough for the cause.

An exchange and a base station are installed at the taluka or county where fibre is located. This exchange functions at a temperature of 55 degree centigrade, and does not require air conditioning. The total power requirement is 1 KW. This capability counters the problem of lack of power in small towns of India. Also, in a situation when power in unavailable, a one KM generator can be easily obtained and used as a backup. CorDECT is capable of offering simultaneous voice and Internet access and can deliver 35/70kbps connectivity to villages that are within a radius of 25 km from the fibre-connected taluka.

The start-up cost for this technology is low. Last year, $200 million worth of this technology was sold and deployed in India and other developing nations including Brazil, Argentina, Nigeria, Tunisia, South Africa and Singapore. An upgrade of this technology will be launched soon. The next generation corDECT technology, which will be released in the latter part of the year, aims to deliver 80/150kbps-sustained rate on each Internet connection.

Such wireless technologies work wonders for 85 per cent of the villages of India, which lie predominantly in the flat areas. In case of rough terrain, primarily mountainous regions and forests, fibre fails to go deep and the problem gets complicated. The TeNeT group is working on a solution that combines satellite and terrestrial wireless to provide low cost connectivity even in such villages. Hence, technology is not a serious issue.

### Business Model for Rural ICT

However, technology is only the first leg. The second leg is a business model, which allows this kind of set-up to scale up to 6, 00,000 villages. The issue that needs to be addressed is in villages where affordability is low, is can a business scale? Telecom operators have declared rural connectivity an unviable business.

The clue to a successful business comes from what was done in the mid 80's in India. During that time, urban area telephony was very difficult. People had to endure a 7-8 year waiting period to acquire a telephone connection. This was particularly difficult on the lower middle class and poor people.

At that time, an innovative idea was developed - find a shop in every street in an urban area and convert it into an operator-assisted telephone booth or a PCO. The PCOs were set up at street corners at a distance of about 50 metres from the closest residential areas and were manned by an operator who kept it open for 16 hrs a day, 365 days a year. The presence of these PCOs addressed the issue of distance and no one was required to undertake a long journey to avail of their services. Such PCOs spread rapidly.

The approach made connectivity viable and pervasive while it also created a stream of entrepreneurs. The success of the PCO revolution can be gauged by the fact that until recently, 25 per cent of India's telecomm income came from these PCOs. Today, 300 million people who do not have a telephone in their house, use these PCOs. The lessons learnt from the PCO revolution were several - aggregation of demand, presence of entrepreneur-driven business, proximity to a facility for greater access.

These three factors provide the basis of a viable business model for rural ICT. Demand aggregation would address issues of affordability, while the entrepreneur and the easy access would ensure a steady stream of users.

**Services**

*Communication*: While telephony is a technology most people comprehend, e-mail and video mail are being offered by the kiosks as additional means of communication, primarily for the purpose of being in constant touch with relatives/friends living abroad or in far away places. Video mail is more popular in the rural areas since the villagers feel more comfortable with a face-to-face dialogue. These technologies are also affordable and n-logue with the aid of TeNeT has created some relevant technologies which function at low bit rates.

*Education*: Education is one of the key applications that is very important to the rural populace. The education module created by nLogue is based on computer-based education and computer education.

*Computer Education*: n-logue has helped create a number of customised courses under the brand name of Chiraag to train people to use computers. Titled Red, Blue and Green, these courses are designed for different ages and skill sets. While the course content of some span over 20 hours, others are some 50 hours long. The course is taught over several classes and each class enables the rural populace to get familiar with computer basics. These packages have proved to be extremely popular in rural areas, as they are seen as means to towards employment.

*Remote tutorials*: The online tutorial is specifically focused on assisting children complete their school examinations. Unfortunately, there aren't too many good teachers at the village level and the remote tutorial attempts to bridge this gap through three modes - learn, practice and test. The remote tutorial is based on a question and answer format. It looks at questions from past papers and suggests the kind of answers that need to be provided. Each module is supplemented with a voice-over that makes learning easy. The online tutorial has focused on several subjects, including Mathematics, Social Science, English and Science.

*Spoken English*: Villagers view Spoken English as an important credential in acquiring a job. Having understood this significance, Spoken English module has been specially created to help children and adults improve their English speaking ability.

*Typing*: Several children come to the kiosk to learn basic typing. They pay a dollar a month and attend 8-10 classes for this purpose. Similarly, adults avail themselves of the kiosk's facilities to acquire typing skills, which they again find is useful in obtaining employment. Apart from this, villagers also may get their resumes made.

*Photography*: Most kiosks double as photography studios, where photographs are shot using a digital camera and a printout can be made available. Photography is proving to be a major source of income. Photographs for a variety of purposes, including for government forms and veterinary applications, are being shot at the kiosk.

*Others*: Children make greeting cards using the computer, which is surprising considering that their knowledge about computers is only some six months old.

*Agricultural Consultancy and vetenary*: Farmers in the village bring their animals to the kiosk and get them treated by doctors remotely. A videoconference is initiated between the farmer and the doctor. Conferences are also welcomed among farmers. Through such conferences, queries, doubts and apprehensions get resolved. These services are extremely important to the rural people as their livelihood depends on agriculture and cattle/poultry.

A current example is the case of the farmers of T Ulagapichampatti. Their okra produce was turning yellow. A videoconference between the farmer and agriculture specialists, in the city, was set up. The leaves and the produce of the damaged crop were shown through the web-cam, the kind and amount of fertilisers added was also discussed. The experts diagnosed it to be yellow mosaic. Appropriate treatment was administered and the farmers were able to prevent a loss of Rs 1,40,000.

nLogue is also planning to enhance the services provided to the farmers through videoconferencing by accessing a greater number of resources and through additional services like weather reports, crop price and other market related information.

*Health Care*: Even though Public Health Centres exist in every three to four villages, many of these centres function rather poorly. Health care is a major concern in rural areas. Qualified doctors in reasonable numbers are available only in towns.

In May 2002, an enterprising kiosk operator took the initiative to send some pictures of an elderly woman's eyes to n-Logue. The lady was suffering from severe eye pain. The pictures were forwarded to doctors in the Aravind Eye Hospital, a large facility in a neighbouring city.

The doctors used Internet-based video-conferencing to examine the patient. This provided the basis for a programme of video-conferencing based eye-care for rural areas. Eye-doctors in town now regularly examine the eyes of patients in the villages by means of video and suggest remedies as might be required. This process was quickly extended to regular health care. Doctors in towns use video-conferencing to provide medical advice to remote patients located in villages.

A multi-party video-conferencing product is commonly used, where multiple villages are simultaneously connected to a doctor in the town. The doctor connects to several villages at the same time and examines them in full public view. Questions of privacy have been raised on account of such a procedure. However, most villagers prefer this mode of examination, and when asked, one remarked that such a system makes the doctors more accountable.

*e-governance*: Similarly, the Internet is being used to approach the government for all kinds of problems. There is a well-known story concerning a handicapped person who sent a complaint to the Chief Minster's cell complaining of unfair termination of service. Having received the complaint, swift action was initiated wherein, the person in question was handed back his job.

3

# ICTs in Rural Marketing

Handheld devices offer the opportunity to extend kiosk-based services as well as handheld-specific services to a broader customer base both within and surrounding the village—thus expanding the market reach of the kiosk. In order to assess the value of these handheld services, constructed a model identifying the success criteria for generating a positive net present value within 5 years.

Based on the following assumptions, the model indicates that handhelds make a successful addition to existing kiosks if they can generate transactions from 7-8 new households per week:

- No additional labour costs would be required to transport the handheld around the village and deliver the service. This is based on assumption that village kiosks are run by families with underutilised labour capacity.
- Handhelds can generate a total of $0.65 per household from new services such as photos and voice mail.
- Handhelds can replicate 50% of existing kiosk-based services and corresponding revenue—$0.30 per household.
- The handheld operator will to travel around the village to deliver the services for $10/week.

- 10% hurdle rate for the initial capital investment required.

As prices of handheld devices continue to decrease according to Moore's law, the one-time and ongoing costs will decrease, and the model will only become more attractive. Based on these assumptions and data points, an investment in one handheld for each kiosk to penetrate more deeply into the rural market environment. The fixed cost of the equipment is taken as per present cost which is expected to drop in future. For example, a handheld PDA company in India, Simputer, is planning to sell handheld at $250-300.

## E-Business Platform for Indian Agriculture Market

About two-third of Indian population is engaged in agriculture. Agriculture being the major source of livelihood in India especially in villages is ironically very backward in technology. The farmers get low price for their products because of limited number of buyers and sellers due to limitations in form of geographical and temporal barriers, lack of price information and lack of technological knowledge.

The e-business breaks these barriers because trading is fast and can be done from any place and with anyone irrespective of his location. It can also provide the latest price information. A negotiable warehouse receipt is issued by a warehouse operator to a person depositing goods in the warehouse. It is accepted by commercial banks as collateral security for grant of loan against the goods stored in the warehouses and a warehouse receipt is also transferable by buying and selling.

But there are certain problems associated with the present system of paper based warehouse receipts like the physical delivery of receipts, which is the final stage of transaction, can lead to extra transaction cost (handling, transporting and storing certificates) and may also lead to losses in form of theft, loss in transit or counterfeiting. An electronic warehouse receipt avoids these contingencies by keeping the electronic record of the ownership of receipts. This supplements e-business as the buyer can be sure about the quality and the quantity of the commodity mentioned in the offer and certificates are not required to be physically produced.

This concept is very similar to the concept of 'dematerialisation of shares' or 'demat' in short, which was introduced in India for the first time by National Securities Depository Limited (NSDL) in 1996. It has succeeded in persuading companies to dematerialise a portion of their shares and allow electronic trade and settlement. So, to make EWR a success there has to be a central body like NSDL.

Presently there are many agriculture e-business platforms existing in the market. Today's world, becoming more competitive every day, demands from business platforms, the flexibility to adjust themselves to the needs and the situation of the market and the targeted section of users.

Some basic features of a business platform for agriculture, which are essential for its success in India, although some of these are general features which apply to any business platform in or outside India:

- *Availability and reachability*: The traders and the farmers, who are the targeted users of this business platform, usually don't have access to internet in the villages and the mandis. In order to ensure the availability and reachability, the medium chosen should be a popular one. Therefore, we have developed the IVR system, which makes the services available through telephones, which are nowadays very common in mandis as well as villages. Also, the number of mobile phone users has increased by

5 lakhs in U.P., the test zone for the digital mandi project (MLA), during the last one year. So, we have also made some services available through sms. The users can perform the price queries and can get the alerts and updates through sms.

- *Language ease*: A large fraction of users i.e. farmers and the traders are not comfortable with English.
- *Computer literacy*: Again, a lot of users are not computer literate either. This problem is tackled by IVR system and the sms center, use of which doesn't demands computer literacy.
- *Reliability*: Although many agriculture e-business platforms exist in the market but they don't ensure that the commodity is actually held by the seller and the quantity and quality are consistent with what is mentioned in the offer. To solve this problem we have come up with the concept of Electronic Warehouse Receipt (EWR) which is the same as the concept of 'dematerialisation of shares' or 'demat' in short. In the model the commodity will be held in the warehouse and the offer can only be posted for that commodity. So, the interested buyers can be sure that the quantity and quality of commodity are what is mentioned in the offer.
- *Related information*: The latest prices of the agricultural commodities are available for each mandi. Also, the digital mandi project (MLA) website, provides the information about crop diseases, seeds, fertilisers, pesticides and the latest developments in the field of agriculture.

## Architecture

- The backend consists of MySQL database engine and Apache tomcat application server running jsp and servlets.

- Clients can connect through conventional web browsers and IP phones. JDBC (Java™ Database Connectivity) has been used for connection with the database.
- As mentioned earlier trading service is also available through telephone network by the use of IVR. For this purpose we have used a voice gateway 'Telepro' from PrologixSoft.
- Various updates and events like new counteroffer or acceptance of your offer are available through sms alerts.

## Web Based Trading

Web based trading provides farmers and traders a forum to buy or sell their products maximising their profits. The major difference from other e-trading sites is that the portal also facilitates fractional trading. A typical transaction procedure goes through following steps.

- A trader places an offer for a particular product along with a negotiable price tag and the quantity.
- Other interested traders may now request specifying the price they would like to buy at and the quantity.
- These requests have three kinds of statuses; waiting, accepted, rejected and confirmed. Initially, all the requests have waiting status which can be changed by the seller to either rejected or accepted.
- After a request is accepted, the buyer has to confirm for the transaction to complete.

**Interfaces**

Any registered vendor is authorised for the following interfaces

- Post a new Offer: can post a new offer for selling agricultural product he/she owns

— Current Offers: can view all the offers currently running
— Search for Offers: search for offers for any particular product
— My Offers: List of offers posted by him/her along with information about the counter requests
— My Requests: List of requests made by him/her along with their status
— My Committed Transactions: List of all the transactions that he/she has committed though portal.
— Vendor Detail: Details about some vendor
— Offer Detail: Detailed information about any particular offer.
— Requests for my Offers: List of all the counter requests for any particular offers posted by him/her. A facility for accepting/rejecting and auto-advisory is also provided here.

## Interactive Voice Response (IVR) System

Interactive Voice Response (IVR) systems provide information either through recorded voice or using a Text to Speech (TTS) converter. The user usually gives input in the form of dual tone multifrequency (DTMF) signaling. When a user makes a call, an IVR application is used to prompt the caller to enter a specific type of information. After playing the voice prompt, the IVR application collects the predetermined number of touch tones (digit collection) and forwards the collected digits to a server.

The IVR application (or script) is a voice application designed to handle calls on a voice gateway, which is a router that is equipped with VoIP features and capabilities. The IVR feature allows an IVR script to be used during call processing. The scripts interact with the

IVR software to perform the various functions. Typically, IVR scripts contain both executable files and audio files that interact with the system software. These IVR scripts are usually VoiceXML documents which may be static or generated by an application server on the fly.

A SIP Gateway from PrologixSoft which picks up VoiceXML documents and processes it. The VoiceXML documents are generated through php scripts running on an Apache http server. When a user dials up he/she is asked to choose a language and then whether to proceed to price-information or to trading. A user can then navigate easily either to obtain price information or to participate in the web-trading. He/She can post a offer as well as reply to some offer from the system.

## Electronic Warehouse Receipt

In present system of paper warehouse receipt negotiable warehouse receipt is issued by a warehouse operator to a person depositing goods in the warehouse. It is accepted by the commercial banks as collateral security for grant of loan against the goods stored in the warehouses and a warehouse receipt is also transferable by buying and selling.

### Present System

There are certain inherent problems associated with the present system of paper based warehouse receipt. Some of the problems are:

- Physical delivery of receipts, which is the final stage of transaction, can lead to extra transaction cost (handling, transporting and storing certificates) and may also lead to losses in form of theft, loss in transit or counterfeiting.
- Warehouse receipt can be used as a security for taking loan or borrow money from the banks. In

case the receipt gets lost, a fresh duplicate receipt is issued by the same warehouse which issued the original one. But this method gives rise to malpractices and forgery. The person holding an original copy which is no longer valid can borrow money or take loans from the banks showing them the original copy.

— There is also a geographical barrier as the receipt is to be physically shown. So, the person can only trade with other local traders.

**Advantages**

— EWR is secure from fraud and forgery. As the information resides on the central database, no person can make a duplicate copy or use an outdated receipt.

— It also breaks all the geographical barriers as the physical production and exchange of document is not required. The seller can transfer the EWR to the buyer and transaction completes when the buyer confirms it.

— Warehouses that use EWRs will not have to print, store, deliver, receive, or stamp paper warehouse receipts.

— Digital Warehouse Receipts are different from paper warehouse receipts in that any part can be fractionalised to thousandths of the whole. And because of the digital nature of a DWR, this fractionalisation can be executed by the bearer on demand.

— Using EWRs help better winning the trust of banks, depositors, lenders, producers and merchants due to its security from fraud and forgery.

— EWR supplements e-business as the buyer can be sure about the quality and the quantity of the

**Transactions**

Some transactions can only be performed by certain type of users (warehouses, users, banks) and types of transactions allowed are different for each.

Every warehouse has a login to connect to the central database and can do following transactions:

- *Create electronic warehouse receipt*: A new electronic warehouse receipt can be issued by the operator if some user brings some products to deposit. The receipt can be issued by entering the product details like grade, price, weight, date of deposit, owners id etc in the database.
- *Handle withdrawal requests*: Warehouse can accept or reject the withdrawal request for the commodity accounting for any EWR. The request is rejeced in case the EWR is on lease or doesn't belong to the user who has made the request.

Every user also has a login for authentication. A user can initiate the following transactions:

- *Selling the product*: The holder of the EWR can transfer a part of his deposited product or the entire deposited product to another user by posting an offer by web based trading.
- *Buying the product*: A user can also place his counteroffer against a selling offer. If the seller accepts, the buyer has to confirm the offer.
- *Submit withdrawal request*: User can also submit a request for the withdrawal of his deposited product. The request is accepted by the warehouse if the concerned EWR belongs to the user and is not on lease.
- *Taking credit from bank*: The user can take bank credit or loan by putting his EWR on lease. Warehouses usually value the price of the commodity at the time

of depositing. The individual banks can allow loans or credits up to some percentage of the value of EWR according to their policies.

Every registered bank is also issued one or more logins. Banks have to handle the lease requests and the release requests. Bank places collateral on the product when the holder takes some loan or credit. Bank releases the collateral when the holder reimburses the amount taken as credit. The user has to make a request for placing or releasing the collateral.

## Mandi Administration

Every registered mandi is issued an administrator login at the time of registration. The administrator can further issue administrator logins and roles within his mandi domain. Every admin. can be issued a role to perform and can have different privileges.

### Some Privileges

Every admin. login in the mandi domain has some privileges associated with it which decide that action can be performed by the user of that login. The privileges are decided at the time of issuing of the login and can be changed by an administrator in the domain of that mandi, if he himself enjoys the privilege to issue logins and roles. The various privileges are:

- *Update price*: The login with this privilege can be used to update the daily prices for that mandi.
- *Update product list*: The login with this privilege can be used to update the list of commodities traded in that mandi.
- *Issue new logins and roles*: This privilege gives the right to issue new logins and roles. Also the privileges of the existing logins can be changed and any login can be cancelled.

- *Privilege to perform warehouse operations*: The privilege gives the right to perform various warehouse operations like issue new warehouse receipt, handling withdrawal requests etc.

**Interfaces**

The following actions can be performed by a mandi administrator provided he has the privilege to perform the action.

- *Update daily prices*: The prices for the commodities available in the mandi can be updated.
- *Update list of commodities*: The list of commodities traded in the mandi can be updated. Also list of the commodities for which warehouse facility is available can be updated.
- *Issue new logins and roles*: New logins and roles can be issued in the mandi domain.
- The privileges of the existing logins can be changed and logins can also be cancelled.
- A new warehouse receipt can be issued if the administrator has the privilege to perform warehouse operations.
- The withdrawal requests for warehouse receipt can also be handled if the privilege to perform the warehouse operations is available.
- Also the latest prices for any mandi and product can be viewed, which requires no privileges and can be viewed by any ordinary user.

*SMS Center*

There has been an explosive growth in the mobile telecom sector. The mobile subscriber base (GSM and CDMA) reached 31.4m at the end of February 2004, with as many as 1.6m subscribers added in February. According to provisional forecasts made by Gartner, cellular connections will reach 56m by the end of 2004. One of the biggest successes in mobile services has been SMS text messaging. According to Gartner, outbound SMS volumes grew 200% to reach 7.39bn messages in 2003. The main driver of growth is the price difference between voice calls and SMS. Most of the vendors in the local mandi own a mobile handset. Infact, the penetration of mobile has been beyond any expectations even in rural areas and cannot be compared to the reach of internet.

**Properties**

The desired properties of SMS center are

- Sends a given message to a given phone number with optional UNICODE encoding.
- Checks regularly for incoming messages and replies automatically to them. This particular module should have easy and high extensibility. New set of request and responses need to be added from time to time as need arises.

**Architecture**

We have connected a Nokia GSM mobile on the serial port of a machine. This GSM mobile accepts the standard AT Commands.

The whole system is essentially a daemon which runs two different threads.

- The first thread i.e., the network. Thread runs a server. The purpose of the thread is to receive TCP/IP requests for sending messages. On receiving a message it "connects to the mobile and sends the SMS.
- The second thread checks the mobile every minute for any incoming message and replies them according to predefined functions.

Two independent threads try to access a serial device which can lead to incorrect sequence of commands being sent to the mobile. Therefore, the main thread maintains a semaphore to prevent collisions over the device. There are two methods inside SMSC class, (a) send SMS, which essentially sends all the SMS's, and (b) check SMS, which checks for new messages every minute and replies them. Both these methods access the device in different modes. Therefore, access to device cannot simultaneously be given to both the methods. The two methods have been declared as synchronised which essentially guarantees that only one thread can be inside only on of the methods at a time.

## North Eastern Region Rural Development

The North Eastern region, comprising of the States of Assam, Manipur, Meghalaya, Nagaland, and Tripura and the Union Territories of Arunachal Pradesh and Mizoram, does not, fall into a separate category of backwardness. Much of the region will be covered under one or the other category of fundamental backwardness. Specifically three types of fundamental backwardness are found in the region viz. areas of tribal concentration, hill areas and chronically flood affected areas.

The North Eastern Region has certain. special socio-cultural features which need to be taken into account in development plans. Considering the fact that the total tribal population of the area is only 4.35 million it is clear that there are a large number of relatively small tribes. Many of the tribes cut across not merely state boundaries, but also international boundaries. Thus, there is a great deal of heterogeneity of population within the area.

The tribal areas do not of course face the multi-caste problems which may be found in other parts of India since there is a great deal of equality within the tribals.

But even here, the distinction between the ruling families and the others as well as the existence, in the past, of slavery in some cases has to be noted. Nevertheless, on balance group conflicts within communities may not be a major problem. What is probably more important is the-possibility of conflict between tribals with respect to land rights etc. which may be more prevalent because of the very large number of tribes in the areas.

The tribals communities of the N. E. differ in significant way from the tribal communities in the rest of India. They are educationally more advanced. For example, the literacy rate of scheduled tribes in the North East relative to the position in India is as follows:

**Table 1. Population Densities**

| State/Union Territory | Males | Females |
|---|---|---|
| Arunachal Pradesh | 34.62 | 17.16 |
| Assam | 8.72 | 1.70 |
| Manipur | 38.64 | 18.87 |
| Meghalaya | 30.11 | 22.79 |
| Mizoram | Included in Assam | |
| Nagaland | 30.17 | 17.68 |
| Tripura | 23.60 | 6.04 |
| All India | 17.63 | 4.85 |

As this table shows the rates of literacy amongst the tribal population in this region arel higher than the all-India average for tribals except in Arunachal Pradesh. The ability of the tribal community to benefit from the job opportunities generated through development process is substantially greater hi the- northeast than in the other tribal areas. There is a stock of educated youth who, if trained, can take-up positions in development, (administration, industrial projects etc.

The tribal societies of the N. E, have lived in relative isolation for a very long period of time and many of the traditional attitudes of tribal communities are still prevalent there. An important instance of this could be the noncompetitiveness of the tribal society where each person has a certain way of living and a standard of life which he cannot change at the expense of other members in the group.

Production for self-consumption and subsistence is the basis for economic activity and calculations based on market values are seldom the basis for decisions on what and how to produce. In this situation when the system is opened to market forces the local tribals, even though they may be well educated, may not in fact benefit.

A substantial proportion of the non-agricultural enterprises in hill and tribal areas, particularly in the towns, are controlled by nontribals. The nominal ownership may well rest with a tribal, if nothing else because of the exemption from income-tax that this would entail. But the bulk of the profits would accrue to the real owner.

This situation obtains not merely in industry but also in trade, transport operations and other services. One reason for this is the inability of tribal entrepreneur to cope with the superior business acumen and enterprise of the non-tribal. The educated tribal, who could perhaps enter these areas of self-employment at present to prefer the more secure avenues of salaried employment open to him. Thus, the processes of non-agricultural development generated in this area do not always benefit the tribal except in the limited area of government employment.

The growth of the market economy in these areas has progressed very much and many communities who lived by barter have been exposed to the cash nexus. Since the trading intermediaries in this market economy are often non-tribals, there is a social gulf between the trader and the customer which opens up possibilities of exploitation. In many ways the impact of the opening up of the economy to market forces is very similar to what was observed in other parts of the country in the nineteen century. Traditional crafts are under pressure as the cost of factory made articles is less.

Indebtedness to money-lenders increases as the cash requirements of the household rise with the decline in traditional crafts and a shift away from subsistence cultivation. The situation of course, varies greatly from one tribal groups to another. There are some tribal groups which have had long tradition of contact with the market economy and they are in a better position to cope with the opening up of their economies.

There are, however, many other groups which have lived a more isolated existence who may well lose out if they are exposed to uncontrolled market forces. These variations in the ability of different tribal communities to cope with a new economy have to be taken into account in any development strategy.

Some of the tribal groups in this region have a matriarchal system of inheritance. There are many versions of this. But broadly speaking, they involve the passing of all property through the female branch. However, the effective management of these properties does not always rest with the women, though the high participation rates of women in the work force are high. In such cases it is possible that the men who work the property may not have a sufficient incentive to undertake improvements and innovations. This is not a matter on which one can give any firm opinion.

## Labour Situation

The labour situation in the north-eastern region is substantially different from what it is in most other parts of India. According to the Sixth Five Year Plan the daily status employment rates in the north-east in 1977-78 were as shown in the next page. As this shows the problem of und]er employment and un-employment is not of any great significance except perhaps in Tripura.

**Employment rate by daily status**

| State/Union Territory | |
|---|---|
| Arunachal Pradesh | 0.35 |
| Assam | 1.81 |
| Manipur | 2.00 |
| Meghalaya | 0.41 |
| Mizoram | N.A |
| Nagaland | 1.03 |
| Tripura | 5.04 |
| All India | 8.18 |

The second distinctive characteristic of the labour force in the region is the high participation rates of women in the labour force in the hill states. The high participation rates and the low rates of unemployment mean that there is very little employment slack to be taken up in the hill states. Because of this it has not always been possible to find labour for construction work in the region even at relatively high wages. But it may be noted that this may also be due to some extent to lack of organisation for mobilising labour.

The position with regard to the educated unemployed cannot be assessed with any firm data. However, in order to indicate a broad picture we present below some date on the ratio of job seekers on the live register of employment exchanges to the enrolment in the relevant stage of education. Job-seekers above matriculate but below graduate level are compared with enrolment in secondary schools and job-seekers who are graduates and above with enrolment in colleges/ universities.

The description of the labour force and the employment situation brings out one important point. Development plans in the hill areas of the north-east cannot be based on the assumption of an employment slack and the alternative employment generated by new developments in horticulture, plantations, animal husbandry, etc. must offer fairly high wages or returns per manday if they are to succeed. The possibility of shortage of candidates for technical and non-technical posts in development administration and in projects must also be kept in mind.

## Agriculture

The north-east is basically an agricultural society. At the all India level the percentage of workers in agriculture and allied sectors was around 72 per cent in the 1971 Census. In the north-east the percentage is high in all States/ UTs other than Manipur, the average for the region being 77 per cent. The hilly terrain of the north-eastern region and the pattern of land availability has led to two distinct systems of agricultures (i) settled agriculture in the plains, valleys and gentler slopes and (ii) jhum (slash and burn) cultivation elsewhere. In jhum cultivation a plot of ground is cleared for cultivation, used for one year and then left fallow. In the following year another plot of ground is tilled and so on till after a number of years the cultivation cycle returns to the first plot. Typically jhum cultivation is practised by hill tribals.

The problem of jhum cultivation is found in many hill and tribal areas of the country. However, the severity of the problem is much greater in the north-eastern region. It has been estimated that around 27 lakh hectares are subject to jhuming in the north-eastern region of which around 4.5 lakh hectares is sown at any one point of time.

This may be compared with the total not sown area of around 33 lakh hectares. The problem is even worse if we exclude areas of settled cultivation like Assam and Tripura. If this is done the area subject to jhuming is around 19.8 lakh hectare, the area sown at one point of time 3.6 lakh hectares and the corresponding net area sown around 6.2 lakh hectares. Thus outside Assam and Tripura more than half the net area sown may consist of jhum plots.

Jhum cultivation is generally conducted without the use of a plough and with the use of the primitive implements. Since the system is subsistance oriented, mixed cropping is common and irrigation is naturally nonexistent. Such a system, may have been viable at the time when population was stable but with growing population the jhum cycle shortens and natural processes of regeneration are cut short. This leads to a loss of soil and erosion and declining fertility.

**Land Tenure and Cultivation**

The system of land tenure and local government in North-East has certain special features which arise to a large extent from prevailing agricultural practices. There are many variations amongst the different communities and the description that follows is based upon papers prepared for the proposed seminar on the Development problems of the North-East.

In most of the hilly regions of the northeast the prevailing method of cultivation of jhuming i.e. slash and burn cultivation. In this system a certain area of jungle is cleared, the debris burnt and the crops grown for one season on the cleared soil. In the next season and for some years there after jhumed area is left and another area is cleared and used for cultivation.

In due course, after say 10-15 years the jhumias return to the first post and one jhum cycle is complete and the second one starts. With this method cultivator shifts from area to area and does not till one plot of land continuously. Hence individual rights to a piece of land cannot get established. The general practice in jhum areas is that each village has a well-defined range of operation in which the jhum cycle as well as other activity like hunting and wood cutting are confined.

Within this area land is allotted to each household on the basis of its capacity or its need, by a village level authority like a village council or a chief, guided by village elders. However, thereafter the land so allotted is not cultivated jointly but separately by each household, though in some cases, operations like land cleaning are undertaken jointly. The location of the settlement itself may change over the jhum cycle.

The village council's generally involve all household heads on an equal footing. However, there is often a distinction between the village elders and the other household heads. The powers of privileges of the chief or the headman vary from case to case. Since the areas within which the village carries on the jhum cycle has to be defined and inter-village territorial disputes resolved there is often an arrangement for a supra-village authority institutionalised in the form of a kingship or a council. Thus many of the tribal areas have a traditional political institution at areas level.

The close connection between jhum cultivation and communal control over and and ownership is attested to by the fact that at least for the tribal groups like the Dimasa, Cacharis such communal ownership does not apply to settled lands under wet or dry cultivation even in jhum villages. These are generally held under individual ownership. The point is further corroborated by the case of the Apa Tanis of Arunachal Pradesh who have settled valley agricultural with individual ownership.

Apart from village lands and lands under individual ownership in certain cases there are clan lands used for the land tenure system in jhum area characterised by communal ownership but operational management is done separately by households. Jhum cultivation, particularly in the land cleaning stage requires a great deal of labour. At the same time the predominent household farm seems to be nuclear and arrangements akin to the joint family system seem to be rare.

The main reason for this is that laws of decent and inheritence seem to inhibit the formation of household. For example, in the case of parmogentive is common and therefore children not in the line descent from separate household. However, with assistance from the parent household if necessary these separate households also become cultivators in their own right. Thus the dominant pattern is that of smaller households with a one or two able bodied adult males. This system when combined with the general absence of joint cultivation leads to some difficulties in organising jhum cultivation on individual plots.

The pattern of land tenure and modalities of land allotment pose several problems when it comes to replacing jhuming by settled cultivation. The development of settled agriculture will require substantial investments in terracing, land shaping, irrigation, etc. The individual cultivator who is to do this must have some assurance that the fruits of his labour will continue to accrue to him and his dependents.

This assurance can only be provided if he is given a permanent title to this land. Since it appears that the principle of individual ownership is generally accepted for land under settled cultivation there may not be much difficulty in this if the distinct and village councils can be persuaded to adopt a progressive policy. A cultivator who gives up jhuming and fakes to settle agriculture may lose his rights in village lands (other than what he is using). He may well be reluctant to do this.

This suggests that the transformation of jhum into settled agriculture has to take up for the whole village simultaneously. This will also help to solve the problem of the land in jhum cycle that is no longer required for crop husbandry. These lands could be converted to village forests, plantation, orchards or pastures. The hilly terrain of the north-eastern region offers ample scope for the development of horticulture and plantations. Fruit production is fairly widespread in the region and at present the area under major fruit crops is about 71 thousand hectares which is a little over 2 per cent of the net sown area. In the case of plantations, tea is of course well established in Assam. However, other plantation crops like coffee and rubber are not very extensive, the area under coffee being 1,984 hectares and under rubber 1,250 hectares at present.

Animal husbandry is an important part of the traditional agricultural system. However, the pattern of animal husbandry in hill areas shows a significant difference from the plains and from the general all India pattern. In these areas the number of pigs and poultry in relation to population is far higher than the all India norm. Thus, excluding Assam and Tripura, the number of pigs par 100 population ranges from 11 in Mizoram to as much as 40 in Nagaland.

The corresponding all India average is around one. In the case of poultry, the range in these five States/ UTs is from 89 per 100 population in Meghalaya to 272 per 100 population in Mizoram. The all-India average in this case is around 25. These high levels are not observed in the case of cattle where the number per 100 population are broadly comparable to the all-India figure except in Mizoram and Nagaland where they are much lower.

The animal holding pattern seen in these hill areas reflects the prevalence of jhum agriculture which does not involve ploughing and therefore requires less by way of animal draught power. In some areas there are social inhibitions against milk consumption which in any case is of less importance from a dietary point of view because of the high level of meat consumption. However, there are signs of change. The pressure of urban demand has led to some growth of milk production and many of the inhibitions about milk consumption are disappearing.

The main orientation of the system, however, is towards meat production. Here too the dependence is on pigs and poultry which can live on waste materials and crop residues so that intensive cultivation of feed /fodder or animal care is not required. However, despite the high levels of pigs and poultry relative to population, there are substantial imports of animals for meat from the plains.

### Non-agricultural Sectors and an Infrastructure

In the traditional non-agricultural sectors the dominant activities are handloom weaving, sericulture and handicrafts. The region has around 13.6 lakh handloom, most of them being looms used mainly for household use rather than for the market. In the field of sericulture, the region has all the four varieties of silk, Mulberry, Muga, Eri and Oak Tasar. Handicrafts based on locally available materials like cane and bamboo, wood, conch shells, etc., are also fairly well spread.

In the modern sector the level of development is limited. There are only about 62 medium and large industrial units in this region of which 48 are in Assam. The total number of small scale units in the region was only 2653 in 1971. The low level of industrial development in the region is a consequence of a variety of factors. The problem does not lie in any lack of natural resources. As pointed out earlier, the region is exceptionally rich in these.

The difficulties rise from the isolation of the region from national markets which alone can provide the scale of demand required for the optimum exploitation of the material and forest resources of the region. In addition there are further difficulties posed by the lack of entrepreneurship, the shortage of technical personnel and a traditional social structure that cannot find a place for the competitiveness and risk taking required in industry.

As far as railway facilities are concerned the Brahmaputra Valley is linked to the rest of the country via a meter-gauge line. There is nominal railway kilometrage in Nagaland and Tripura, but the other constituent units have no rail lines whatever. However, projects to convert the existing meter-gauge line to broadgauge up to Gauhati and to extend the rail network into all the constituent units are underway.

With regard to roads the kilometrage of motorable roads per km. of area is below the national average in all the units except Assam and Tripura and substantially so in the case of Arunachal Pradesh, Meghalaya and Mizoram. There is also a further problem of the inadequacy of links between the north and south banks of the Brahmaputra which is being corrected to some extent through new bridges across the Brahmaputra near Tezpur and Togighopa.

As far as power supply is concerned the hydel resources of the region are very large relative to requirements. The principal problem arises from the nature of the terrain which makes it difficult to deliver power to all points in the region. The foundation for any programme of agricultural development will be on an effective base for basic research and adaptive research and field trials.

At present the research support is provided by the Assam Agricultural University and the ICAR Research Complex for the North-Eastern Region. However, facilities for localised adaptive research are limited. Technology has to be developed on an area specific basis and facilities for adaptive research will have to be put up by the States. The Committee would lay the utmost stress on this aspect and suggest that the State Governments take on greater responsibilities for adaptive research.

In the hill areas the dominant problem is the widespread prevalence of jhum cultivation. As pointed out earlier it is estimated that as much as 27 lakh hectares are subject to jhuming in the hill areas of the north-east. With the growth in population the jhum cycle is becoming shorter leading to a gradual deterioration in the quality of forest cover in the hills.

Apart from this the low productivity of jhum agriculture is leading to a growing imbalance between food demand and availability. The promotion of a shift to settled cultivation in jhum areas must be a central element in the development strategy in this region.

Jhum cultivation was the response of the tribal community to a situation of abundant forest land, limited labour and the lack of flat lands suitable for food and fibre crops. The logic of jhum cultivation also led to a system in which the control of land was not vested with the individual but with a village council or chief. There have been several schemes to control and transfer jhum cultivation. They have emphasised a shift to plantatiou or horticulture or terraced cultivation.

It has been estimated that at the end of the Fifth Plan about 60 thousand hectares of jhum land and about 25 thousand families have been changed to settled cultivation under the various jhum control programme taken up through the NEC or under the State Plans. This amounts to barely 2 per cent of the area liable to jhuming and 6 per cent of the jhumia families. Hence the programme of jhum control has barely begun.

The problem of jhum cultivation is so widespread that all the households involved in jhum cultivation cannot shift to settled cultivation even in a 5-10 year framework. Hence the strategy for jhum control must distinguish between different areas on the basis of the extent to which the jhum cycle has become short and ecological problems created thereby. On the basis of surveys, jhum areas should be put into three categories :-

(i) Areas where the cycle is still above say, 10 years and where ecological problems are not yet acute. In these areas at present the emphasis should be on improving the productivity of jhum through better agronomic practices, land conservation measures and improved varieties ;

(ii) Areas where the cycle has fallen below say 5 years and where ecological problems are already acute : These areas should be taken up on a priority basis for conservation to settled cultivation ; and

(iii) Other areas: These areas will probably face acute problems in the near future. Hence, in this case, the immediate measures may be for improving jhum but there should also be a programme for the gradual introduction of settled cultivation.

A variety of models of settled cultivation relevant for jhum areas are available : -

(i) *Conservation to plantations*—This is the earliest of the proposed strategies, and involves a total shift to a plantation, crop like tea or rubber.

(ii) *Terraced cultivation*—This is being actively pursued in State Sector schemes and is being implemented with and without irrigation.

(iii) *Conversion to grassland*—This is happening in certain areas where animal husbandry and dairying has developed.

(vi) *ICAR's three-tier systems*—This involves crop husbandry with bench terracing in the lowest third of the slope, horticultural crops with half moon terracing in the midportion and forestry in the top third.

**Valley Agriculture**

In the case of the chronically flood affected plains the strategy must be to reorient the cropping pattern and crop varieties to suit the flood regime. The dominant crop in these areas is paddy. The bulk of these areas are in Assam when 500-600 thousand hectares are planted under autumn paddy (sown in early June and harvested in August-September) and 1500-1600 thousand hectares under winter paddy (sown in June-July and harvested in December-January). It may be noted that while yields from autumn paddy are of the order of 700kg of rice /ha. the yields from winter paddy are around 1100 kg. of rice/ ha. However, in berth cases yields have not shown any growth trend during the post Fifth Plan period.

The other major paddy growing area is Tripura which has roughly 130-135 thousand hectares under autumn and winter paddy and Manipur which has about 35 thousand hectares under autumn paddy and 140 thousand hectares under winter paddy. In these areas, unlike the Assam, the difference in yield between the two seasons is less marked. In Tripura the yield figures for autumn and winter paddy being 1050 kg. of rice /ha. and in Manipur they are of the order of 1600 kg. of rice /ha. for both seasons.

The primary problems of paddy cultivation in this region is of adjustment to the flood regime in the flood plains in the Brahmaputra basin. Paddy once it is established, can survive submergence upto 7 days or so. The real problem is to design a strategy for those areas where the period of submergence can be expected to exceed this. In such areas the Committee has suggested two approaches:

(i) The paddy varieties sown in the monsoon season must have a long duration so that they mature after the last flood;

(ii) Cropping patterns should be adjusted to avoid the flood season and use the surface moisture, supplemented by irrigation for rabi and summer crops. In this connection, the programme to promote rabi wheat in rice growing areas is relevant. Both these strategies are applicable in the flocd affected areas of the Brahmaputra basin.

The extent of irrigation in this region is limited. Whatever irrigation is available is used for paddy. In 1976-77, 23 per cent of the paddy area in Assam, 7 per cent in Tripura and 42 per cent in Manipur was under irrigation. In all these states the potential for surface and ground water irrigation is quite substantial. The development of this potential is vital if the strategy of avoiding the flood season is to be implemented.

One important irrigation possibility in the area close to the river is the low cost shallow tube well. In the case of small farmers the utility of this option has been questioned on the ground that their holdings are highly fragmented. In such cases water can be lifted from the river, brought to a contour channel at some height and released to the fields below. Such a system would, of course, have to be in the public sector.

The low level of crop production technology in this region is reflected in the very low levels of fertiliser consumption which is of the order of 2-3 kg/ha, in Assam and Tripura, though it is somewhat higher in Manipur where it is around 12 kg/ ha. The modernisation of agriculture in these valley lands will clearly require a massive increase in the use of fertilisers. Fertiliser consumption can only increase if leaching can be avoided. Hence the strategy of avoiding the .flood season is important Blue-green algae systems and biofertiliser can provide a suitable source of nitrogen in water logged areas.

**Horticulture**

The region is suitable for both tropical and temperate fruits and horticulture can provide a very lucrative alternative to jhuming. However, the present pattern of agronomic practices will need to be changed. The development of horticulture will require a major extension effort to improve agronomic practices and place them on a more scientific and commercial basis. The success is extending new fruit crops like apples in Arunachal Pradesh suggests that such an effort can be mounted. The extension effort will have to be backed up by the provision of nurseries to supply suitable planting material and research to identify promising cultivator. However, there is a wide gap on the field effort.

There is hardly any follow up after the plants are distributed resulting in heavy mortality and low productivity. This gap in the extension effort will have to be covered if horticulture development is to prosper. In order to do this major programme of technical personnel will also be needed. The second major gap in the field is the lack of effective marketing support.

In many cases marketing problems arise from the lack of communication facilities. Hence the horticulture programme must include the formation of growers cooperatives which would be supported by the proposed regional corporations for agricultural marketing. These facilities. It may also be necessary in the first instance to concentrate in areas which are connected by road so that communication problems are minimised.

### Plantations

The three principal plantation crops of relevance for the north-eastern region are tea, coffee and rubber. Of these the first two are relevant principally for hill areas and the last one for the plains. The extension of plantations in hill areas can only be done slowly as the rural population is weaned away from low productivity jhum farming. A more rapid pace of development will lead to shortages of labour for working in plantations. In migration of plantation labour from outside of the sort which took place when Assam tea plantations were developed is, today, not acceptable to the people of the region. In fact, because of this possibility of immigration, State Governments appear to be reluctant to pursue this particular option.

The plantation development programme has to be based on the small holder approach. Compact areas can be taken up for development by a corporation with each individual family being given the right to about 2 acres. In the first stage the compact area would be developed by the corporation, with the individual being employed as labourers.

Once the plants reach the bearing stage, the management would rest with the individual family who would have access to common facilities and technical advice from the corporation which would also take on the responsibility for marketing. The individual family would have the right of usufruct but would not have the right to alienate the land or convert it to some other use. Such an approach will help to ensure that plantation development takes place in an economically viable manner and that the benefits accrue to local people.

**Forestry**

Even though the North-Eastern Region has abundant forest resources, these are not being exploited except in small pockets. Limited or restricted control of the State Government/ U.T. Administration on the forest areas, lack of population, lack of trained manpower, communication bottlenecks, absence of processing industries etc., are the factors standing in the way of efficient exploitation of the forest resources. The rich forest resources of the northeast cannot be exploited unless the pre-investment survey identifies areas of low, medium and high stock so that priorities for afforestation and exploitation can be worked out for different areas.

It will also be necessary to extend the silvicultural control of the Forest Department to forests outside the reserved /protected category. The problem of roads will have to be taken care of by building in the required provision in commercial forestry schemes. Forest policy in this region will also have to take into account other activities like tasar culture and horticulture in drawing up plans for development.

**Animal Husbandry**

It has been pointed out earlier that the animal husbandry system in the hill areas of the north-east is oriented towards the meat production. Hence the emphasis should be on pigs and poultry. However, in the plains and near urban areas in the hills milk-oriented developments are taking place. In many ways the north-east is well-suited for animal husbandry. The high and well distributed rainfall provides excellent conditions for the development of grass lands which can remain green throughout the year. The region is full of nature forage consisting of perennial and annual plants. Meat consumption in the region is high. Despite this the hill areas of the region import substantial quantities of meat from outside.

The foundation of the animal husbandry programme is the upgradation of animal productivity through crossbreeding. The high altitudes in this region can allow high levels of exotic blood. In the case of cattle the upgradation programme will have to rely on frozen semen based artificial insemination. In the case of pigs besides pure bred exotics, cross bred boards may have to be used for infusion of 25% exotic blood so as to ensure wider coverage.

Poultry development in the form of intensive poultry farms or upgradation of backyard poultry will also have to be based on cross breeding. The facilities required for producing the breeding animals are now in position in the form of regional breeding stations. Frozen semen is available at Khanapara. Hence the problem now is that of delivery.

With regard to feed requirements, pigs and poultry rely to a large extent on wastes, crop residue and local vegetation. However, prepared poultry feed for the interior poultry farms will be necessary for which purpose suitable feed factories will be required. In the case of cattle, range type farming based on grass lands and forage trees many of which have a high nutritive value may be the answer. The upgradation of meat and milk productivity will not be achieved unless a suitable marketing system is established. In the case of milk the system of growers cooperatives linked to urban milk supply scheme provides a frame work.

However, in the case of meat and eggs no such framework exists. Since the role of meat and eggs production is very substantial in the animal husbandry system of the north-east, marketing supply in this area is vital. The proposed regional exploitation for agricultural marketing can provide this support at the apex of the marketing system. What is required is a village level organisation akin to the milk cooperatives. Unless this is done the bulk of the animal husbandry system may remain subsistence oriented and there may be little incentive to raise productivity through crossbreeding or better feed management.

**Sericulture and Handloom**

Sericulture is a traditional activity in the north-eastern region and comprises ericulture, mugaculture, oak tassar culture and mulberry culture. The problem and the measures required differ for each of these varieties. Ericulture based on castor plantations is a low return activity and is conducted to a large extent for self consumption. The main problem lies in the fact that worm is reared on castor leaves. New plantations have to be created every year which involves considerable recurring expenditure. It is, therefore, necessary to extend the use of substitute plants like Alitanthus glan-dulosa or excelsa which is quick growing perennial species along with this improved spinning devices are necessary to raise productivity and earning per unit of time.

Mugaculture is also indigenous to' the region. Muga seeds are procured from hill tracts though the worms are reared in the plains. Because muga worms deteriorate in the plains the whole operation has to be started a new every year. Muga plants take five years to grow and a vast area of high land is required to cultivate them. The returns to the family from muga rearing are not very high because of low productivity at all stages from the plantation to the laying to the reeling.

Any improvement in mugaculture must depend on the measures to raise productivity like intensive rearing of quality seed sources, preservation in cold storage of seed worms, avoidance of the hazardous rearing period etc. Mulberry culture is also traditional to the region. However, because the culture is age old, there is no standard breed of silkworm in the region, and the worms produced are of inferior quality. Moreover, the plantation are often of mixed varieties of mulberry which can affect yarn quality adversely.

Therefore, the crucial elements in the strategy have to be the rearing the high silk yielding variety of silk worm race and improvement of leaf yields by extension of high yielding varieties and appropriate agronomic practices. Oak tassar culture is a more recent introduction. However, so far it has caught only in Manipur. If the labour used is essentially unemployed or under-employed the return from oak tassar culture are fairly attractive. With improvements in yield rates and sex ratio the returns can be raised further. Since oak tassar offers an important avenue for settling jhum as it is necessary to pursue these productivity improvement measures.

A general problem affecting all area of sericulture is the low productivity at the reeling stage and the exploitation by middlemen. Hence the extension of improved spinning devices as well as the provision of raw material and marketing support is necessary. In this respect the Group Centre approach described in the National Committee's Report on Village & Cottage Industries along with the district level supply and marketing society recommended in that same report can provide a suitable organisational framework.

**Industry and Minerals**

The limited development of the modern industrial sector has been noted earlier. This is reflected in the fact that in the recommendations of the National Committee on the Development of Backward Areas presented in the Report on Industrial Dispersal the whole of the north-east is included in the areas that should be eligible for incentives for industrial dispersal.

The types of industries that can be promoted in the north-eastern region fall broadly within the following categories: -

(i) Major raw material based industries, which in the north-east would be mainly paper, cement and petro-chemicals.

(ii) Industries to supply local demands, where the scale of local requirements is large enough to sustain an economically viable unit.

(iii) A variety of small industries falling in category (ii) above or based on agro-processing e.g., fruit canning, meat processing, timber processing etc.

The essential task for planning at this stage is to identify the potential for development, as has been done for cement and paper, and to promote these industries in a manner which will maximise local impact. At the present stage of development of entrepreneurship and technical skills the impact of small industries based on local markets or on agro-processing may be greater. Hence industrial promotion must be directed to these types of industries rather than to large capital intensive projects.

It is essential that industrial development in this region should involve local people, to the maximum possible extent. In the case of large industry dependence on entrepreneurs from outside the region or on the public may be unavoidable in the immediate future. However, in the field of small and medium industries, local entrepreneurs should be promoted vigorously. This will require an effective programme of entrepreneurial development and a support system for small industry.

### Transport Development

Many development opportunities in the north-east cannot be exploited because of the lack of transport facilities. Hence investments in this sector are, to a large extent a precondition for effectively implementing not merely industrial development programmes but also market oriented programmes in horticulture, plantation development and village industries.

The theoretical transport requirements of the region for optimum development are so vast that they cannot be met for a long time. At the present stage of development transport development in the north-eastern region would have to be based on the specific needs of each project and programme. Moreover, given the large gap in requirements, every attempt must be made to locate projects and programmes in manner that minimise the need for additional infrastructure.

Thus market oriented horticulture can be developed fast in areas already served by roads. Forest based industries can be located in forest areas which can be readily opened for exploitation. Despite this further major investments in transport will be required and should be included in the plans of the States, the NEC and the Centre. The resources available for transport development can be stretched further if standard of construction are recommended in the light of likely traffic to identify possible economies. The different components in the Transport system viz., railways, roads, inland water transport, air transport have to be developed as an integrated system. Piecemeal improvements or extension of network may not serve much purpose if they do not fit in with the total capabilities of the system at all points.

# 4

# Cottage Industries

A marketing network can be created with the help of governmental agency, cooperative or a voluntary association for collective action. There are certain inherent difficulties in having collective action on voluntary basis on a sizable scale. It presupposes mutual understanding and appreciation on the part of the participant for its success.

The absence of such factors is found in abundance, and hence doubts about the workability of the voluntary scheme can well be understood. In the scheme of things envisaged, it is proposed that the marketing functions would be undertaken by the covering organisation which would also be responsible for performing other functions. Institutions undertaking only marketing functions may not be very effective in all situations.

The marketing of village and cottage industry products has to reflect modern trends. The bulk of the output of village and cottage industries consists of consumer goods many of which compete with similar products produced in the organised sector e.g. textiles, matches, food products, footwears, leather products, etc. Hence the marketing system for village and cottage industries must be equally organised end sophisticated.

The first requirement is that the village and cottage sector must produce a product which in design, quality and process can compete with organised sector output. It is true that some cottage industries products, in the field of textiles and handicrafts in particular, have certain uniqueness and traditional designs which are their selling points. However, even for these products some product development is necessary e.g. for furnishing fabrics in the textiles sector or new types of products in brassware.

Hence continuous monitoring of consumer preferences, fashion changes and design developments in the organised sector is necessary. A related area is that of packaging which has to be sufficiently attractive to compete for consumer attention in the shop. Some of these things have been done e.g. for exports markets by the Handloom and Handicraft Export Corporation. However, the responsibility for these aspects is not clearly focused. It cannot be done by the individual artisan or even a district level organisation.

Therefore it must be the responsibility of the state and central level set up for promotion of village and cottage industries and of the marketing organisation. The second requirement is that the product must be available where and when the consumer normally purchases it. This means that retail outlets cannot be the same for all village and cottage industry products. Khadi bhandars and similar set ups have a role to play. But it is too much to expect the consumer to hold back on his purchases of a variety of items like matches, honey, footwear, textiles till he visits the bhandar. Each product must be available at the retail outlet at which it is usually purchased by the consumer. The marketing organisation must

(a) identify the appropriate outlet for each product,

(b) arrange for the product to be sold through this outlet,

(c) send sales representatives regularly to make supplies, provide promotional material and service the retailer.

In many cases the marketing organisation may find it more economic to operate through distributors in the private retail trade appointed on a commission basis to discharge these functions. This is how the large consumer goods producers in the organised sector operate. The third major requirement is that of product promotion at local, regional and national level. This requires that for each product a strong brand image is built UP and projected through advertising in mass media and other promotional measures.

Since the output will be produced by a mass of artisans in different parts of the country the brand image must be that of the state or central marketing organisation. A producer based brand image will be useful only in— exceptional cases. The responsibility for such promotion must clearly rest with the state and central level marketing organisation. The fourth major requirement is the need for a strong link between the marketing and production organisations.

This is necessary to ensure that adequate stocks of the product are available where and when they are needed. Product promotion will be useless unless the product is produced in adequate quantities and in time and despatched to where the demand exists. The mass marketing approach outlined above constitute a significant departure from present practice in which, with a few exceptions, there is heavy dependence on a few outlets like the Khadi Bhandars and Super Bazar.

These outlets will continue to play a role and the super bazars in particular can help in the marketing of a fairly large amount of "cottage industry output. But these retail outlets have a limited potential and a quantum jump in cottage industry output is only possible if the products of this sector compete with organised sector products in all retail markets, particularly, those in urban areas. That this can be done is proved by the experience of Amul Butter and Lijjat Papad, to give two examples. These two examples highlight the workability of the scheme in different situations. In case of butter Gujarat Dairy Development Federation organises production at centralised points even though the collection of milk is from a large number of small producers.

Quality of the production is ensured through quality control measures and the butter, thus produced, is marketed under brand names through the network of retailers. In case of Lijjat Pappad, the production is through a large number of workers. These workers get their raw materials from centralised points and their product is taken back after strict adherence to quality control measures. The final product is well publicised, like in case of butter, and sold through the network of retailers. Most of the marketing organisations limit their area of operation so as to ensure better returns. This system leads to exploitation and malpractices and calls for rectification.

It is very desirable that the major chunk of production is distributed through a large number of outlets and the private retail channels are optimally utilised. To give a boost to marketing, a link up between the supply of raw materials and marketing should be maintained. With the assured supply of raw materials, on reasonable terms, the artisan needs to be assured of a certain amount of off-take at the minimum. It is very difficult to fix any definite proportion as the nature of products is such a composite lot. Broadly, it may be assumed that an assurance of lifting 50% of the product from the artisan would give him great relief.

The balance of the 50% may be left to him to meet his own personal needs and for marketing to the demand of his clients in the local markets at hand. The certain cases, like that of carpet manufacturing, a hundred percent lifting of the product may be desirable as the local demand may be negligible. As such, it would be desirable to specify the limits of off-take of products, in each line of production, after taking due cognizance of the magnitude of the local demand. In any case it should be ensured that the value of the off take shall at least cover the cost of raw material supplied,

Identification of markets, other than the local, would obviously have to be entrusted to some agency. At the district level the function can be rightly undertaken, by the DSMS (District Supply and Marketing Society). A similar sort of function for the state level and international level could perhaps be best performed by suitable state level and national level institutions respectively. The proposed agency at the State level viz. State Rural Industries Development Corporation would be the right agency to undertake functions at that level.

The pilot programme of rural marketing centres has taken upon itself a very wide franchise. In the present concept the RMC is also not supported by any higher level expertise. The Committee is of the view that the pilot project as at present envisaged cannot work for lack of necessary expertise in the RMC to manage all the items in its franchise.

The Committee has elsewhere recommended a separate raw material and marketing organisation to cover village and cottage industries. For technical support and training, other organisations of a hierarchial nature starting from the group unit have also been recommended. If these recommendations are accepted,- the role of the rural marketing centre will be purely one of maintaining a suitable display and marketing centres at the urban level for all types of village industries.

The committee would recommend that the operation of the RMC may be limited to this particularised service for the village and cottage industry. An existing marketing centre under any of the All India organisations like KVIC, Handloom Board etc. has already developed its expertise for dealing with the products of that organisation.

One need not tinker with the structure already built as far as the product of that parent organisation is concerned. However, the Committee would recommend that al! these structures should become multi-disciplinary centres and sell such other products which can be profitably handled. But for the products of other organisations and other group units covered by the marketing chain, the rural marketing centres should preferably act as agents for sale. In effect, the rural marketing centres may work as an 'adathiya'.

A suitable commission arrangement will have to be made and the parent production organisation will also have to make provision for taking back unsold goods and marketing them elsewhere. As this may seem to indicate that the rural marketing centre has no role to play in pushing the wares of the village industries other than that of the parent organisation, the Committee wants to make it clear that these must be promoted as effective sales centres.

At the same time, in order to give a incentive to the unit for effectively pushing goods of industries other than that of the parent industry, suitable incentives should be built into the sale system. As the Committee has recommended that a district level society (DSMS) will be incharge of the raw material and marketing cover, that organisation will have to look after all these aspects of effective marketing.

The responsibility for maintaining quality of goods, studying the consumer market and putting in the right type of goods at the urban centres should be the responsibility of the DSMS where intensive development is contemplated.

Where new rural marketing centres have to be developed, it is desirable that from the beginning they are developed as multi-commodity display and sales centres. They should be placed squarely under the DSMS. Thereby the spread of urban marketing will be controlled by the society in an effective manner. At the same time the Committee wants to make it clear that one sale centre in an urban market may not be enough for pushing the entire production of village industries to market. Existing private retail shops should in any case be fully brought in the chain of distributors.

The marketing organisation has to cover one very important problem of the present artisans marketing system which is that he has to continuously produce and earn a livelihood whereas his wares have generally a seasonal market. Even the local buyer of his goods prefers to link up his purchases of the village artisans' goods with the festivals when it is traditionally utilised.

The artisan at present not being able to hold back his goods has to go to the middleman, trader who does the effective distribution for the festival seasons. The marketing organisation will have to take over this important role. The organisation should be in a position to take over stocks of the producer on an agreed production level throughout the year. Whilst stocks may be displayed in the urban markets through the marketing centre and retail traders who are willing, a special push will have to be done during the festival seasons.

In these seasons, the goods will fetch a higher price and larger sales can be done. The strategy of marketing will have to allow for all this. This means that the strategy of marketing cover must have facilities for storage to keep goods in condition till the festival season and push the wares in the ready markets. Incidentally, the festivals for various village industry goods are different in different regions of the country.

The overall marketing cover given by the State Rural Industries Development Corporation and the Central corporation will have to do this services for exploiting the festival markets of the country effectively. At the district level there has to be storage accommodation for the society and necessary financial support for holding back stocks. A suitable linkage with Government purchases can yield an assured market for the products of the village and cottage industries.

Government agencies are meant to include all Central and State Governments, Defence, Railways, P&T and other public sec tor undertakings. In Tamil Nadu reservation of purchases from village and cottage industries has been made obligatory, and the scheme is reported to have shown encouraging results.

For Governmental purchases specifications and designs need to be settled in advance and should be adhered to. Ad-hoc change of specifications can put the industry in an awkward situation. Here in the example of blanket purchases by the Defence may be worth mentioning. A sudden change in the specifications of the lot ordered led to huge stocks with KVIC which found it very difficult to dispose off subsequently. For ensuring regular supply to Governmental agencies a proper tie up of the basic production units with districts and State level agencies and operation of a national grid for marketing needs to be ensured.

The Committee has already suggested that DSMS should take back from the artisan at least 50% of his goods if not more. Presently, wherever marketing of goods of the artisans has been attempted on this basis, like in the Handloom Cooperatives and in the Coir Board, it has been found that there is no emphasis on ensuring quality of goods. It should be made clear that the artisans produce will be accepted by the marketing organisation only if it fulfils the basic standards laid down by the organisation. There should be no compromise in this matter. Whilst placing orders on the artisan, the marketing unit will have to continuously survey the market and see that the demand is for saleable varieties and standards.

Thus a continuous appraisal of the market and its needs and for ensuring the quality of the goods to meet the standard will be the responsibility of the DSMS. Whilst this may be theoretically a sound proposition, in actual practice there can be two problems. Firstly, pressures can be brought in on the organisation to accept goods below quality.

Secondly, having a sort of monopoly cover over the artisans ware, bad practices may develop in the marketing society so as not to pay a proper price for the goods by unnecessarily rejecting goods even though up to quality. For both these reasons it is desirable that a neutral body should be constituted where disputes of this nature can be taken up arid arbitration secured. The Committee suggests that such a neutral body may be set up by DIG for each district comprising of technical experts who will be relevant to the problem.

The decisions of these technical groups will be binding on both parties. The present marketing organisation of the KVIC or the Handloom Board generally proceeds on the basis of cost plus. Such a method firstly does not look into the legitimate costing of the goods and secondly does not fully tap the consumer market. The marketing organisation should, therefore, see to the proper pricing of its ware. In its price to the artisan it should ensure a reasonable daily wage for work done. At the other end the price of the goods should be what the market will bear. In practice working of this system will show that in the festival seasons, price can be suitably marked up provided that standards are maintained and proper marketing efforts made.

This opportunity to make profits should not be lost by the society. On the other hand, in the off-season, goods may not be saleable except at the cost basis. To push the goods and clear the goods quickly from stocks and enthuse the artisan to produce more, the strategy should be for a lower price during the off-season. Unfortunately, our entire approach has been contrary. We maintain a high price in the off season and a subsidised price in the festival seasons. This is not what the trade does. These niceties of trade practices will have to be built into the marketing structure.

The standard specification for goods and the organisation for checking the quality has to be built into the marketing system. There has to be a continuous appraisal of market demand. Soon after the Handloom Committee made its report in 1974, when the Development Commissioner was appointed in the Ministry of Industrial Development, research was carried out about the change is taste of people regarding handloom cloth for the general population.

It was found that contrary to the existing thinking the demand was moving up into higher counts of yarn. There is a continuous process of taste change in the consumer and greater and greater sophistication in the demand. There has to be some organisation which studies the market patterns and changing tastes and then prepares the specifications for new types of goods that have to be produced.

This responsibility can be spread between the Central Government and the State Governments. The district marketing organisation and the State marketing organisation can draw the standards for purchase of goods from the artisans on their assessment of the market.

The whole focus to the development of village and cottage industries need to be market oriented and commercial and not sheer production-oriented, and should ensure fair wage to the artisan. In the production-oriented approach the prime focus remains confined to providing social benefits at the cost of the State /Central Exchequer. Such an approach can only bring us near to an untenable situation and should not be encouraged for any length of time.

A market-oriented commercial approach, on the other hand, would reduce the dependence of the village and cottage industries. The art of selling consumer goods lies in effective advertisement. The general experience shows that the relative price of goods and the relative quality and lasting nature of goods has many a time no relevance in the consumers opting for buying a particular product. That is why advertisement has come to stay as an important component of the marketing effort.

Village and cottage industries have not been able to follow this effective marketing strategy because each production unit is too small and getting together sufficient numbers to support an effective advertisement drive is an impossible task. The Committee recommends that in the initial stages the marketing through advertisement will have to be a service to be rendered by the State organisations for the development of village and cottage industries. A subsidised service for this purpose will be fully justified. This of course will have to be linked up with the capacity of the chain of organisations to produce the quality and quantity required in the market.

Handloom experience shows us clearly thus if a proper marketing strategy is evolved, it should be possible to sell the products of village and cottage industries in new and distant markets in the country. Cooptex of Tamil Nadu and the parallel organisation in Andhra Pradesh (APCO) have shown how by organised marketing the consumer can be tempted to favour these goods.. The States will have to take the initial lead in developing such covering organisations for the industries most prevalent in the State.

The Centre's contribution will be provision of market intelligence, training institutions, design development and suitable guidance to the States in improving their marketing strategies. Many products of the village and cottage industries are finding special markets in foreign countries and Committee recommends full tapping of such potential markets. Today this sector of high manual involvement has a market in the affluent countries as handicrafts.

Though our- Handicrafts Board has done pioneering work in this direction much more can possibly be done in this and other fields. The skilled worker and the artisan will find their mettle only in meeting the taste of this sophisticated market. Incidentally return for manual will be the highest in such a market. The credit aspects, in their entirety, are being dealt with by the Reserve Bank or India's Committee to Review Arrangements for Institutional Credit for Agricultural and Rural Development. This Committee, headed by Shri B. Sivaraman has submitted its report.

The recommendations of this Committee would be examined to find out whether it has suggested suitable safeguards in respect of the problem at hand. Credit requirements for marketing of production to the artisans will be fairly substantial. A rough estimate made by the Gaya Organisation, the Committee has already referred to, puts the demand between Rs. 30 and 50 lakhs per year. Even though the ways and means credit can be obtained from the commercial banks or the cooperative system, margin money will be required for both the DSMS and SRIDC. Sufficient margin money will have to be provided by the State Government to these two organisations based on a reasonable appraisal of the business that will be developed. Adjustment of margin money should also be made from time to time as business develops.

The Reserve bank of India has been providing credit at cheap rates to the handloom sector which works through cooperative societies and cooperative marketing organisations. It is only reasonable to expect that similar cheap credit facilities will be given for marketing of production to all the other village industries that are taken up for development by the States provided they are in the cooperative sector. But the problem is that the number of artisans that have to be brought within the frame of development is so large and the Committee has already indicated that it is too much to expect formation of viable and efficient cooperatives to cover this vast number of artisans within any reasonable period.

In fact, the Committee is of the view that expecting a cooperative structure could develop in any large measure in the near future will be unwise. The DSMS and the SRIDC has been postulated by the Committee in order to meet this gap in organisation to this sector. The cheapness of the credit is not really meant for the cooperative but is to enable the cooperative to give cheap credit to the artisans. The Committee recommends that the DSMS and the SRIDC who have to provide marketing facilities to the artisans at cheap rates should be enabled by the Reserve Bank to get its ways and means credit for this purpose at cheap rates equal to those given to the cooperative system.

The Central Government has already indicated that a Bill for formation of a National Bank for Agriculture and Rural Development will be introduced shortly in Parliament. When such an organisation comes into being, it should be possible for that organisation to support both the cooperatives and the State Corporations on par for the necessary funds for supporting artisan's classes.

Till then, some method will have to be found to give cheap credit to the DSMS and the SRIDC. The Committee recommends that this should be considered on a priority basis and an answer found early by the Central Government.

The average artisan uses rather outmoded equipment. In most cases, even though better equipments having higher levels of productivity and returns are available, the same are not being made use of. This raises the cost of production to a level which the market cannot absorb. To steer clear of such a situation efforts need to be directed towards increasing the productivity of the artisans. A three pronged attack on the problem of technological development of the artisans is required.

In the first step an effort must be made to equip the artisan with the better techniques of production that are already available. The second stage would be to remove the drudgery element from these techniques, through technological development, so as to make them more productive. The third element would be to innovate new technologies through research and development activities.

Since the inception of planning there has been constant endeavour to highlight the research and development aspects of cottage and village industries. In the process some developments have taken place. However still much remains to be done.

The First Five Year Plan recommended measures which sought to enable village and cottage industries to carve operational areas for themselves and also to aid them in tapping the same efficiently. The prime measure suggested was that of the Common Production Programme whose possible strategy elements included, among other things, coordination of research to alleviate the temporarily disadvantageous position of village and cottage industries. In the Second Five Year Plan a wider horizon of rural industrialisation was perceived in the broader context of increasing production of consumer goods.

To fulfil such aims, structural changes to ensure modernisation of rural industries in economic fashion was foreseen through State initiative and assistance. The Pilot Projects of the Second Plan and the Rural Industries Projects under the Third Plan also stressed the research and developmental aspects of rural industrialisation.

Technological development of industries in rural areas was considered sine-qua-non of their success. The Fourth Five Year Plan stressed the need for progressive improvement in the production techniques of small and rural industries so as to enable them to produce quality goods and bring them to a viable level. The technological development programme for traditional rural industries aimed at a two-fold objective viz., the avoidance of technological unemployment during the period of transition and the increase in productivity of these industries through adoption of improved techniques for attaining a viable basis.

In the subsequent Fifth and Sixth Plans evolution of an appropriate technology for village and cottage industries has been suggested so as to remove the drudgery element, increase the earnings of the artisan, reduce the price of products and improve their sale ability through ensuring quality production. The various Industrial Policy Resolutions and Statements have also highlighted these aspects.

The Khadi and Village Industries Commission has been, in a way, the sole agency concerning itself with the advancement of khadi and village industries. It has taken some concrete steps towards research and developmental activities in the context of the rural decentralised sector. The scope of the research activities undertaken at the behest of KVIC include experiments in design, fabrication of improved equipment, modification in the existing implements and improvement of various processes of production. The research activities carried out by the KVIC come under the overall R&D programmes which are examined by the National Committee on Science and Technology (NCST).

The Science and Technology programme of the KVIC was approved by the Government of India in consultation with the Planning Commission in the financial year 1975-76. Initially 9 projects were sanctioned. Subsequently more projects were identified and by the end of 1978-79, 42 projects were under way. All these projects have specific objects to be fulfilled in a specified period. Though the R & D effort is a continuous process it acquired significant attention when N.C.S.T. formulated special panel for KVIC.

With the constitution of a standing Committee for guiding R&D efforts of the KVIC and formulation of panel of experts for village industries under KVIC, these efforts-are now gaining depth and direction. In the scheme of DJCs, due recognition of the development of research and technology has been there and a full fledged manager has been envisaged in the set up to look after the relevant aspects. The functions, as defined, require this manager to keep abreast of R&D in select product lines and quality control method and to ascertain problems of entrepreneurs in quality of raw materials, production methods and processes. In concept a manager has been provided, but in reality hardly any one has been appointed.

A lot of effort has been put in by various organisations to improve the existing technology so as to give an artisan a better return for his labours. Yet, considering the picture as a whole, the effort so far made is haphazard and no strong institutions have been built up yet to identify the problems and find the remedies.

The organisation structure will have to meet these requirements. It is evident that very large numbers of the artisans plying village and cottage industries are at present using equipment which is obsolete whilst much better equipments have been devised and can be made available. The technology followed by them is also not the most productive according to available technology. An organisation has, therefore, to be built up to see that the millions of artisans who have not yet adopted the technology are enabled to take advantage of the available technology. This require an organisational cover to-

(i) train the artisan in the available technology;

(ii) equip this with the necessary equipment to adopt the technology; and

(iii) provide a trouble shooting cover whilst his adoption of the technology is going on.

The cost of equipment for the various village industries, though modest, is well outside the capacity of the average artisan to procure on his own. The KVIC has introduced, system of capital subsidy. Under the industry component of IRD and the TRYSEM a subsidy upto Rs. 30GO/- per artisan has been provided for. The system must ensure that the artisan gets his equipment without undue pressure on his resources. A suitable system of grant and loan has to be devised.

The first task of the technologists will be to find a relevant mechanisation for the drudgery element in the present manual operations in the various village industries. Firstly, the operations involving drudgery will have to be identified in the production chains of all the village industries which we are going to support under the present available technology. This is a technical task and has to be done by the experts, for the industries which the Committee has suggested should be taken up on priority. For each of the groups of industries, technical expertise will have to be brought into effect to analyse the present production chain and identify the points for technical improvement.

As in the view of the Committee this is the most important improvement that has to be Drought into the village and cottage industries sector, this process should be completed within a year of the start of operations. It will be for the technologists to specify the remedies quickly. The remedies for drudgery elements having been found and already in several industries such aids are available; these new aids will have to be built into the production chain. Such aids will be common to a large number of artisans. For example, carding of cotton will support a large number of artisans.

Like that we can identify existing technologies and newly developed technologies for each industry. The common service is best provided at the growth centre unit. If this is too small for the productivity of the common service, a number of group centre units will have be brought within the command of such a service. This will have to be planned for each industry by the technical hierarchy for that industry. Having identified that technology and provided the necessary service for aiding the artisans, the artisan will have to be taught to utilise the service and improve his productivity.

For the same time that he now employs in the traditional production process, by reducing his drudgery a greater opportunity-is given to him to use the time so saved in producing the final goods. Thus, each artisan will be producing larger amounts of the goods under the new dispensation. Unless there is an organisation to take charge of his larger productivity and spread the market, the present system of supply and demand will not enable him to effectively dispose of the surplus. As a result, the improved technology will not lead to improved productivity. The crucial problem here, therefore, is that of marketing larger amount of goods in the economy.

The intermediate technology which will enable mechanisation of certain additional items in the process chain has therefore, to be developed for each of the village industries taken up for development. The technical group for- the industry will have to identify such points where mechanisation can be brought in. Here, there has to be a rapport between the field technical hierarchy and the research institutions to identify the points in the chain where improvement in technology is possible and then proceed to do the necessary research and development to provide the relevant technology and the equipment. Thus, in the pursuit of intermediate technology a rapport has to be developed between the field technical organisation and the research workers.

A close look at any village industry with a view to developing the technology will show that the problem is a multi-disciplinary one. There are very few technological research centres in the country which can handle a multidisciplinary problem. The national laboratories mainly are uni-disciplinary bodies though a few of them have got the capacity in particular sectors to look at more than one discipline. It is only the regional research centres of the CSIR where multidisciplinary problems can at present be handled.

The KVIC is at present having an organisation for research and technology. It is necessary first of all to identify the various national institutes and research and technology development for village and cottage industries. This can only be done at the national level. This postulates a national organisation which can study the basic problem and identify the necessary institutes. Having done this, the problems and suggestions that come up from the field from the various technical groups will have to be analysed and a direction given to the research and development programme in each of the village industries. This again is best done at a national organisation.

Most of the national institutes and research and development institutes today work on a custom service. Their own research, like those of the laboratories of the CSIR though given the franchise to support industries have not yet got down to the level of village and cottage industries, in their own research programmes.

To demand that these institutions should now look into intermediate technology on a priority basis will not work because of the inertia of the system unless some organisation is prepared to act as the customer and pay for the research. The national organisation will have to be one that should perform these services for the village and cottage industries. This again will be the most effective way of quickly utilising the structure of research and technical development already available in the country.

The Department of Science and Technology has already provided a precedent by supporting the KVIC for funding certain accepted programmes of research and development. This cover will have to be extend to the national organisation to meet the needs of the problems as identified by them. Whilst the chain of organisations and the methodology proposed would meet the problem of gradually improving the technology of the artisan to meet his productivity, certain mundane problem arise. The experience in the field of KVIC and others shows that even though the equipment has already been designed which improves productivity of the artisan and the artisan is prepared to accept the equipment and there are provisions for training, there is no structure for producing the new equipment and supplying them in large quantities.

A suitable organisation will, therefore, have to be developed for assessing the demand and farming out the work. If a careful assessment of the demand is made and a time phase of not more than 10 years to cover the entire artisan group in the country under the most important industries is accepted, it will be seen that the task of equipment, production and distribution will be a very substantial one.

The equipment needed by the artisans is possible to be produced at different levels. In some cases the production at the level of individual artisan or a group of artisans may be possible. In others the production may have to be assigned to small scale and medium scale sector of the industry. Yet in some other sectors, like leather technology, the desirability of producing the machines in the large scale sector may be necessary. Some other management problems will also arise in trying to rapidly transform the technology in such a vast number of production units.

The artisan has to be trained for adopting the improved technology. It is desirable to proceed on a well spread pilot project approach taking a group unit under each industry to intensively train all the artisans in the group. With the experience gained, further refinement of the training process can be done and large scale adoption ensured. Another problem is that the technical hierarchy for each of the village industries which are in the field and which it has been proposed should be expended, most of them are not fully aware of the technology improvements that are going on in their sectors. There has to be an organisation for suitable in service training for such technicians and technologists.

Prevalence of outmoded production techniques coupled with low traditional skills is still broadly the order of the day with artisans in the backward areas. In successive Five Year Plans, stress has been laid on imparting suitable training to artisans and others engaged in the village and cottage industries, but the programme implementation thus far does not present a consistent picture of implementation of various schemes in a well knit and coordinated fashion. The First Plan envisaged coordination of training and research.

During the Second Five Year Plan, training aspects became quite conspicuous. Under the RAP this Plan envisaged training of artisans through provision of stipends. Under the Pilot Project scheme the introduction of training-cum-production centres to introduce new skills in rural areas by combining the advantages of institutional training and apprenticeship system was launched. Planning of training facilities and improvement for skill was mentioned as an important element of the Rural Industries Projects under the Third Plan.

Skill development was sought through cluster type institutions for training in allied trades in selected areas, peripatetic demonstration and training in village industries and handicrafts, and all India Institute for training in industrial extension techniques. The Fourth Plan laid stress on training programmes for skill improvement in the selected trades particularly of rural artisans. Through the concept of growth centres, the Fifth Five Year Plan envisaged a carry forward of the need for training. The training programme of the village and cottage industries is primarily handled by the five all India Institutions viz.

(i) KVIC,

(ii) All India Hand-loom Board,

(iii) All India Handicrafts Board,

(iv) Central Silk Board and

(v) Coir Board.

As at present the KVIC comes under the domain of the Ministry of Rural Reconstruction, Coir Board under the

Ministry of Industry and the other Boards under the Ministry of Commerce. Beside these institutions, the State Governments and voluntary agencies are also engaged in the process of imparting training. KVIC has been engaged in the training of artisans and supervisory technical and managerial personnel. The training programme of the Khadi and Village Industries Commission is quite comprehensive and has four tier structure i.e.

(i) training of officers and staff of KVIC and State Boards and institutions,

(ii) training of managers, accountant and supervisory staff of institutions,

(iii) training of unit level workers, and

(iv) training of artisans.

The first tier training is conducted at the Central Training Institute at Nasik. The second tier courses are conducted by zonal training centres and village industries training institutions. There are at present 2 zonal centres located at Nasik and Bangalore and 41 khadi and village Industries training-cum-research institutes spread in various parts of the country. The third tier training is conducted at Khadi Gramodyog vidyalayas which are about 100 in number and located in various parts of the country and also through zonal training units. The fourth tier training is conducted at production centres. There are also Khadi and Gramodyog Vidyalayas in most states which are being run by voluntary institutions.

In the case of khadi with the technical and technological improvement, artisans have been given orientation in the improved techniques at Jamuna Lal Bajaj Central Research Institute at Wardha and various other centres spread throughout the country. A similar sort of programme has been followed in the case of other village industries. In some industries some specialised institutions have been imparting training as well.

For instance, in case of cottage match industry, training is imparted at Kora Kendra, Bori-vili, Bombay and also at Vellore (Tamil Nadu) and at Kurukshetra (Haryana) to candidates deputed by registered institutions, co-operatives and Slate Boards who are willing to set up cottage match units. Stipends and travelling allowances are paid to candidates and tuition fee is paid to the institutions. In case of beekeeping a central research and training institute has been set up at Pune.

A departmental training centre has been set up at Dim a pur in Nagaland with a view to catering to the needs of the North Eastern States Under the expansion programme of New Model Chark-has, an NMC refresher training course has been conducted at Bengeri, Hubli district. Two vidyalayas, one each in Haryana (Punjab) and the other at Dhanan (Haryana) have been set up to meet the needs of the expansion programme of khadi.

Other training programmes have been upgradation of the weavers' training centre at Kumaritkatta (Assam), setting up of a training centre at Yahali (Arunachal Pradesh), under the auspices of the Arunachal Seva Sangh and training of KVI and RAP Managers under the Government's DIG programme at Rajkot (Gujarat). Management oriented training has been organised from time to time in collaboration with institutions like Small Industries Extension Training Institute (SIETI), Hyderabad, Indian Institute of Management Ahmedabad etc. Co-operative training is also arranged through co-operative training colleges managed by National Co-operative Union of India.

At the national level the National Institute of Rural Development has been set up. This Institute has been organising several training programmes as part of its servicing activities to central/state governments and other organisations. Training about industrial co-operatives is also being done by KVIC. Uptill 1978-79, KVIC has trained 7,06,894 persons. Of this, 573,964 were in khadi and the rest in other village industries.

The training programme in the field of hand-loom is mainly carried through (i) the Institute of Handloom Technology located at Salem and Varanasi and (:i) 23 Weavers Service Centres spread all over the country. A new institute of handloom technology for Not h Eastern Region has been envisaged. The expert weavers impart training to existing and prospective weavers. With the starting of schemes like Intensive Handloom Development Projects and Export Production Projects, the training needs of this industry have gone up.

The Central programme for development of Handloom Industry includes training programme for managers and secretaries of co-operative societies and apex bodies. Scheme for education of members of Co-operative Societies has also been launched. The co-operative training programmes are arranged through Vaikunth Mehta Institute of Cooperative Management, Pune, National Council of Co-operative Training, Co-operative Union of India and Institute of Management, Ahmedabad.

The All India Handicrafts Board has launched massive training programmes for training of persons in rural and semi-urban areas in various crafts. The training programmes include both training of craftsmen on large scale and apprenticeship training and are imparted by the training centres. After the training, the artisan can go either for self-employment or work in the production centres.

The Central Silk Board has three Research Institutes located at Mysore, Berhampore and Ranchi and provides three courses namely post-graduate, in-service refresher course and farmers training course. The State Governments also provide training of the level of

Certificate course in sericulture. Training colleges run by the State Governments are in the States of Assam, Bihar, Jammu and Kashmir, Orissa, Karnataka, Tamil Nadu and West Bengal.

The Coir Board has been running a National Coir Training and Design Centre at Alleppey which provides training to the managers as well as artisans. The Centre is running training courses in the field at advanced coir technology and medium term as well as short training courses for artisans. On the spot technical advice and guidance is given to coir processors in coir extraction, spinning, dyeing, shade matching, weaving etc.

Training in coir processing and fabrication of simple coir products is also imparted to local artisans in the demonstration-cum-extension centres. The Coir producing States are also running a few training programmes in their produce-Hon-cum-training centres. Regular training programmes for artisans under the Rural Industries Projects and Rural Artisans Programmes have been merged with the DICs. DIC has accepted the-concept.

The agency of DICs have taken full recognition of the training aspects and one of the managers in the set up is earmarked to deal with this problem along with the research and development aspects. He has to arrange for training courses in production methods and in the management of small and tiny and cottage units.

The programme of National Scheme of Training of Rural Youth for Self-Employment (TRYSEM) has been initiated with effect from 15th August, 1979 with the principal objective of removal of unemployment among youth. The main thrust of the scheme is on equipping rural youth with necessary skills and technology to enable them to take to vocations of self-employment. The training programmes cover all the three sectors of the economy viz, primary, secondary and tertiary.

During the period of training, stipends upto Rs. 100 per month can be granted to each trainee, and a grant upto Rs. 50 per trainee per month is given to the training institution so as to meet the training expenses. A lump sum amount of Rs. 100 per trainee per course is also provided for purchase of raw materials, tools etc.

During the course of training, the trainees are helped to prepare project reports, which are converted into bankable schemes. They are helped to apply for bank loans and subsidies. Subsidies throughout the country are on IRD pattern and have a maximum limit of Rs. 3,000 per trainee. Besides the institutional mode of training. TRYSEM also permits and encourages nonformal training through industrial units, services establishments, master craftsmen, artisans and skilled workers. Strengthening of existing training infrastructure is an important component of TRYSEM.

The entire expenditure on TRYSEM is shared on 50 : 50 basis by the Centre and the States. Concurrent evaluation is an integral part of the TRYSEM scheme. During the course of operation of TRYSEM for a few months, a number of deficiencies have come to notice. There are various areas in the country which have absolutely no training infrastructure. Even eise-where, the rural vocational training infrastructure either serves the needs of urban youth, or is lying in a moribund state for want of resources, staff and equipment. It is suggested that each district in the country should have at least one composite training centre.

New training centres of the composite type need to be set up in remote inaccessible areas, hilly and tribal areas on a priority basis. State Governments find it's exceedingly difficult to make resources available for setting up of new training centres or strengthening of existing ones, under the current pattern of 50:50 sharing. This pattern is also likely to tilt the balance further in favour "of the relatively advanced states, as their financial capacity is better.

It is, therefore, suggested the training of training infrastructure, whether old or new, should be done on 100% basis by the Central Government for centres to be located in remote and inaccessible areas, hilly and tribal areas, and areas predominantly inhabited by members of the scheduled castes. This is the only way we can build at least one well equipped composite training centre in each district of the country to meet the very large demand that the recommendations would entail.

For course of less than one month's duration the stipend should be fixed at Rs. 5 per day. For all training-programmes, the cost of travel to and from the training institution should be allowed to be funded from the scheme. There are variations in the pattern of other agencies in the decentralised sector like the Handicrafts Boards and KVIC. It is difficult to promote training in such items without accepting the pattern already approved for the purpose by the concerned organisations. These patterns have the blessings of the concerned Ministries. It is, therefore, suggested that such pattern should be treated as patterns approved under TRYSEM also.

Expenditure on such trades should be allowed to be incurred according to such pattern. Provision of an improved tool-kit is one of the important factors in raising the productivity of a rural artisan. Most of the Schemes like the Rural Programme or the SFDA do have a provision for a free tool-kit or some subsidy. TRYSEM has no such provision. It is recommended that this crucial element should be built in as integral part of the scheme. There is no difference in the pattern of subsidy whether a project is intended to benefit an individual beneficiary or is of a community nature. In the case of community irrigation schemes, 50% subsidy is already approved under the IRD/SFDA.

A similar provision needs to be made for TRYSEM projects to be launched on a cooperative or community basis. The scheme has hardly any provision for training of trainers or for strengthening the infrastructure relating to trainers' training institutions. This needs to be included. In some areas, no progress can at all be made unless a corps of specialised trainers is first built up. This is a critical input. On the credit side, there is no mechanism to ensure that the banks charge only the differential rate of interest from the TRYSEM trainees. In fact, reports indicate that rates of 9 to 11 % are being charged.

While there has to be an attempt to induce the banks to be more liberal in their approach, it also appears to be necessary to subsidise the interest rate, so that the effective rate charged from the rural artisans does not exceed the DRI interest rates. Many of the official and non-official agencies have brought to our notice that they consider availability of raw materials and organisation of marketing to be the most critical factors, which will be decisive as far as the success or failure of TRYSEM is concerned. At present, the scheme has no provision for these items. It is also felt that there is no single agency which is willing to take on a coordinative, catalytic and financing role in this, regard.

It is, therefore, suggested that there should be appropriate schemes for share capital participation in rural industries marketing corporations, setting up of rural marketing and service centres and improvement of village mandies and haats. There are a large number of voluntary organisations engaged in training programmes for rural development and some of them are doing commendable work. The development of these organisations has not been uniform either chronologically or geographically.

With regard to training of community leaders and village technicians under the Hundred Village Development Programme the report says that 202 locals were trained as bare-foot technicians who manned health posts and operated and repaired pump-sets, maintained accounts, trained people in industries and various other trades like carpentry, tailoring and spinning of all the 202 beneficiaries trained, 59 are working for the Ashram on its staff in the various pockets, 131 are working in their own villages either as counsellors or helpers to the Gram Swarajya Sabha. Five of these people have been sent to other institutions outside the area to assist them. Seven of them were able to find jobs on their own outside the area.

Another experience of training programme through voluntary efforts worth mentioning is that noticed in the Valod area of Surat District in Gujarat. VPSS (Vedchhi Pradesh Seva Samity) is the voluntary agency working in the area since 1954 and its activities have been advanced through the assistance of KVIC. VPSS viewed the purpose of the training programmes as

(i) to introduce skill among the youths for such occupations which would provide them better remuneration on one hand and on, the other cater to the needs of the society,

(ii) to upgrade the skills wherever necessary and

(iii) to introduce new skills and services in view of changing pattern of consumption.

In the third category of training, courses were conducted for prescribed period. In the other two categories, training was given to candidates till they were found proficient. Normally the training period was of six months to one year depending upon the efficiency of the trainee. For the selection of the candidates wide publicity is given in the villages and the selection is primarily confined to candidates belonging to families below the poverty line. Craft or trade selected for the candidates is commensurate with his aptitude, his study and other family background.

VPSS has set up production centres and trainees are attached to these centres. Such centres include garment making, carpentry, hand-made paper, weaving, oil pressing, printing press, poultry management. Trainees are also attached to master artisans working in the areas on their own.

Fields covered include radio mechanic, commercial printing, wood carving, tappers, brick layers. Trainees for certain pursuits were admitted to training institutions located in outside areas including urban areas. The training courses herein include truck driving, automobile repairing course, T. V. technology, air conditioning, wire man's course, turners, mat making, dyeing, printing, hand processing of cereals and pulses, bamboo and cane work, short-hand and typing.

The centres providing training are given tuition fees and lump sum for wastage, per trainee. The training scheme is planned and implemented through the planning centre of VPSS at the Central level. Periodical assessment of the programme and follow up actions for post training period like providing loan to the trainees for setting up his unit, is arranged so as to make the training programmes fruitful.

Purchase of machines, tool kits etc. on wholesale basis is also organised for the trainees. The foregoing has broadly surveyed the training facilities created so far and indicate the inadequacy in relation to the desired level of requirements and reach needed in far flung areas of the country. It has been noted that variations exist with regard to duration of training, rate of stipend, cost of training etc. even within the various All India Boards.

The Committee has already advised development of village industries through a group approach. In the various traditional village industries though, the country has already developed a superior technology with equipment which can double or treble production per unit yet this technology and equipment has not reached lakhs and millions of artisans in the country.

The introduction of existing technology and the best equipment available has to be made to lakhs and lakhs of artisans in the country. Obviously, our usual methods of training centres will not answer to the situation. The Committee, therefore, recommends that the approach should be to select master artisans in the industry who are acquainted with the equipment and the technology to be located in each group under the group approach for training the artisans in the group over a period of time by peripatetic handling.

The Committee would suggest that the approach may be for the master artisan to first of all demonstrate the available technology with the best equipment possible at the growth centre of the group. Having done this initial institution, he should take the artisans in the group in small numbers and train them by practical work at the centre for a week or two. After that, when the artisan has some capacity to use the equipment, he should be provided with the equipment in his own village and at that stage, the master artisan should move round the villages trying to help the artisans in their surroundings to improve their handling of the equipment.

It is only by pilot scheme started immediately in the various industries that norms of training through master artisans can be laid down and followed on an all India pattern. It is, therefore, felt that each of the apex organisation responsible for selected village industries should now pay attention to first of all identify the technology and the equipment that is now available and pursue pilot project of master artisan training of group of artisans and establish the necessary norms.

The introduction of mechanisation to avoid drudgery will have to be on a common basis for those parts of the operation preferably at the group centre or in a number of suitably located centres within the group. The artisans to handle these machines will have to be trained at common training centres. Where new technology has to be introduced with mechanisation of the individual equipment, it is desirable to adopt the master artisan approach, because our ultimate objective is to transfer this technology to all the artisans who are lakhs in number. For this, the master artisans selected on the basis of the existing technology are to be first trained in the new equipment and then used as the guide on a peripatetic basis in the groups selected for the introduction of the new technology.

In the initial stages, this will have to be done on a pilot basis and then the future master artisans for training can be identified within these pilot groups and sent for training to other areas. The group technician will have to act as a support to the master artisan in advising artisans in his group. It is, therefore, necessary that all the group technicians should be suitably trained in the existing technology and further trained in the intermediate technology which may be developed from time to time. This will obviously have to be done at suitable common training centres.

Higher levels of technicians and administrative staff will have to be given management training and familiarised with marketing practices etc. All this will have to be in various training centres which will have to be identified by the DICs with the help of the State authorities. The Committee has already pointed out that traditional industries do not exist in all parts of the country even though environment and the market for the industry may exist almost throughout the country. The present location of artisans, is a matter of historical accident.

If backward areas have to be developed artisans may have to be developed in new areas wherever the need exists or the raw materials exist. For this, the Committee would suggest that the group approach would be the right approach for this training. Wherever the new skill is to be introduced a group of artisans should be selected and their training should be undertaken at a common training-cum-production centre on the lines of the carpet training that is now being given by the Handicrafts Board. The centre where the training was started was on the principle of the training centre being handed over to the trainees for production once they have acquired the skill.

Practical experience has shown that the demand for such artisans in carpet weaving is so much that the training centres continue to be training centres, whilst the trainees are absorbed in various carpet weaving centres by private enterprises. A similar situation can arise where training of artisans will be necessary in the non-traditional areas. The training centres and the use of training-cum-production centres should be ultimately adapted to the requirements of the particular situation.

For every training programme, there is to be expenditure on the staff that does the training, some stipends for the trainees if they have to come out of their houses and some amount for wastage during the period of the training in raw materials or finished goods. The TRYSEM Scheme for training unemployed rural educated youth is suitable for adoption to the training programmes for village industries.

# 5

# Globalisation and Rural Development

The phenomenon of globalisation is far from truly global. Whether measured by increased investment or by greater trade openness, globalisation is restricted to a relatively small number of developing countries. According to a recent World Bank study, about 2 billion people, one third of the world's population live in non-globalising countries, a category which includes large parts of Africa, and many Muslim states.

In contrast, the globalising group within the developing countries, comprises 3 billion people, including China, India, Brazil and the Philippines. While there is evidence suggesting that participation in globalisation reduces poverty, the foregoing discussion of the role of the Green Revolutions suggests that these were equally, if not more, instrumental. In recent years the globalising developing countries have grown faster than the rich countries, and their incidence of poverty has declined. Yet, even for this group, there is presently some doubt about how much momentum there is in the process of globalisation. In the last four years private capital flows to emerging economies have been running at only two-thirds of the level of the previous three years. Thus the mid 1990s were the high point of globalisation, and we cannot be certain that this will be bettered in the present decade.

It is likely that broad-based smallholder development has played a critical supportive in enabling poor countries to successfully engage with globalisation. Among other benefits, success in smallholder development expands domestic markets for consumer goods, stabilises food prices, gives government more confidence to open the economy and provides the means for rural people to invest in schooling.

This secures a platform from which to open up to international trade and investment. If it is accepted that in the globalising poor countries prior success in smallholder development has laid the basis for successful participation in globalisation, and that both processes have been powerful reducers of poverty, then the obvious policy questions are how to extend these processes to the non-globalising poor countries.

An interesting initial point is that simply opening a non-globalised country to investment and trade may not realise the full potential for poverty reduction, as the platform of broad-based smallholder development has not been created. A deeper point, is that for the remaining non-globalising countries, smallholder development may be harder to achieve, for a number of reasons:

- They often exhibit more difficult fundamental agroclimatic and/or political conditions;
- Late entrants to competitive markets experience huge challenges in catching up, as early innovators have accumulated a base from which to continue to forge ahead with technological improvements, while prices for commodities (as opposed to differentiated improved products) are likely to suffer from oversupply. The evidence can be seen in established, and possibly accelerating, long term trends towards declining world prices for agricultural commodities.

- Globalisation depends critically on a 'transactions infrastructure', which comprises not only the obvious aspects of roads, harbours, telecoms etc., but also a basic understanding of the transactions supporting roles of government in setting and maintaining a minimally acceptable policy environment in terms of macroeconomic management, taxes and trade restrictions, and an institutional infrastructure of acceptable legal codes and their enforcement. The reality is that in many of the non-globalising countries, the 'transactions infrastructure' is very weak and sometimes deteriorating further. Addressing this problem is a long-term challenge.
- Trends in the development of supply chains based, as noted above, on exporting to markets with the incomes levels found in the rich countries, are potentially highly exclusionary for the nonglobalisers, as well as for those producers within the globalising poor countries whose footholds in world markets are not well established, and perhaps based simply on commodities. For agriculture, horticulture and floriculture, the activities in which smallholders can engage, the end markets are dominated by large retailing firms, which compete among themselves on continuing minor innovations in products and packaging, on maintaining strict quality criteria and on price. These retailer-dominated supply chains require producers to be able to:
  - meet exacting quality criteria, covering such matters as size, colour, texture, pesticide residues and taste;
  - adjust production volumes rapidly to meet short-term market trends;

- track minor product innovations by changing planting material, planting methods and packaging;
- keep up with cost-reducing technical progress, in a context in which the partner retailer and its competitors have multiple sourcing.

These requirements are enormously demanding in terms of information flows, capital requirements and governance and management of the system. Dispersed smallholder suppliers are at an increasing disadvantage, as they have much greater difficulties in accessing and then acting on rapidly evolving price and technical information. Furthermore, as explained below, financing smallholder agriculture is often very difficult, this being exacerbated by the current policy climate.

Another emerging trend, in which smallholders are likely to struggle to be included, is the 'decommodification' of some traditional commodity industries. This can be seen in the production of, for example, maize and oil seeds bred to have high proportions of the component most valued by the processors, and also consistent quality in this respect. This trend is likely to be accelerated by technologies based on genetic modification.

There is no reason, in principle, why GM technologies cannot be developed to be appropriate to the farming conditions and markets of poorer smallholders. It is a near certainty that the bulk of R&D investment will be concentrated on technologies for highly capitalised commercial farms supplying processors whose quality criteria are both exacting and different from those of traditional smallholder markets.

An important question is the extent to which GM technologies developed for large farmers will be able to spill across to smallholders with only modest modification costs. Many of the GM technologies introduced into smallholder farming may not be particularly tolerant of smallholder conditions, where the full input package may not be affordable, ability to absorb detailed technical instructions may be less, and there is greater reliance on direct rainfall.

Early work on cotton suggests that this may be too pessimistic a view, but it should be borne in mind that this concerns an early GM innovation, i.e., creating insecticidal properties in cotton plants, which are producing exactly the same product, i.e., GM technology is not being used in this case to enable de-commodification. A final aspect of globalisation is that its underlying theory is neoclassical economics.

The neoclassical model, based on certain stylised assumptions, predicts that increased openness to flows of trade and capital will raise welfare in all participating countries. In the case of trade and poor countries, the key mechanism is held to be that countries will export those goods which make intensive use of their most abundant factor of production, and in so doing raise domestic demand for this factor.

Thus the growth of labour intensive exports (such as smallholder produce) will raise wages and returns to self-employment in the supplying country. Similarly, capital flows from richer countries to capital starved poorer countries should put capital to work with abundant supplies of land and labour, giving higher returns than is available through investing in capital rich countries.

The logic of the neoclassical model is powerful and, as noted above, the empirical evidence is broadly consistent in that, in general terms, open developing countries have fared better in growth and poverty reduction than the non-globalisers. However, as is explained below, neoclassical economics ignores transactions costs and the importance of the institutional issues which influence transactions costs. Even in the mainstream economics debates about developed countries this is becoming to be seen as a major weakness.

Recently Hall and Soskice have elaborated a framework which explains comparative advantage among rich countries in terms of their different institutional infrastructures, which are in turn based in their histories. In other words, this is a theory that comparative advantage has institutional foundations, and has little to do with the neoclassical idea of trade patterns being driven by 'factor endowments', except in the most obvious cases. Comparative advantage determines what a country will specialise in, both in supplying export and its domestic markets, unless underlying tendencies are overridden by strong policies (such as the agriculture policies of the European Union).

Hall and Soskice's theory that comparative advantage has an institutional basis has profound implications for poorer countries, although these authors have not yet published their ideas applied to development. However, a simple carrying across of this basic idea to poor countries suggests that weak institutions may deny poor countries comparative advantage where neoclassical logic suggests strongly that it should be present.

The general implication is that a necessary condition for development is the building of transactions-enabling institutions. Within this, the challenge of building institutions to support smallholder development is critical for success in the earlier stages of development, and this seems to be a precondition for successful participation in globalisation.

The policy prescriptions of neoclassical economics are a thoroughgoing liberalisation of external trade and the domestic economy to capture the gains available from trade and capital flows. The economic role of government should be limited to regulation and the supply of public goods and merit goods, both of these arguments tempered by a moderate scepticism based in an appreciation of the possibilities of government failure.

## Smallholder Agriculture

Smallholder agriculture have a long and rich tradition, stretching well back into the 19th century. The insights developed by these traditions by focusing on three aspects of what is new: (a) global trends in technologies and markets; (b) trends in policies; and © fresh analytical insights.

The view that smallholder agriculture did not have key strategic role in development was central to the development economics of the 1950s. The highly influential Lewis model saw smallholder agriculture as often little more than a labour reserve, from which workers could be drawn for the growing industrial and service sectors. Specifically, Lewis assumed that in the earlier stages of development labour could be withdrawn from smallholder agriculture without raising the productivity of those workers who remained.

This led to the infamous notion of a perfectly elastic supply curve for labour, which implied that in the early stages of economic development, industry and non-agricultural services could grow fast, as new workers could be drawn in without raising the wage rate, because the marginal cost of labour in agriculture was unaffected by departures from the sector. For Lewis, smallholder agriculture was implicitly a 'black box', within which low productivity might be due to either extreme overcrowding and/or the weak integration into the market and non-capitalist behaviour within the sector.

The Lewis view of smallholder agriculture was comprehensively challenged by scholarship from the 1960s to 1980s, as discussed below (and for convenience labelled 'the Mellor view', although many authors contributed). But from the later 1990s, a new questioning of the strategic role of smallholder agriculture emerged within what has become know as the 'livelihoods approach' (e.g., Ellis).

Influential strands within the livelihoods literature emphasise the extent of diversity in the income generating activities of the rural poor, and suggest that a prior (Mellorian) 'agricultural fundamentalism' has downplayed diversity and its policy implications. According to this perspective, agriculture may be an important way forward for improving the livelihoods of particular groups of poor people, but it should not be 'strategically privileged', i.e., it should not be seen as necessarily a more effective path out of rural poverty than any other livelihood opportunity that appears to be available to the poor in the local environment.

This view can be summarised in the words of Ellis: 'The recognition of diversification as a widespread strategy within a livelihoods approach to rural poverty necessitates moving away with the previous preoccupation with the small farm as the sole or main platform for rural poverty reduction'. An important aspect of the 'strategic role of agriculture' view is that the promotion of smallholder agriculture is a 'win-win' strategy, because smallholders are an efficient user of resources and also an equitable approach, as it increases returns on assets held by poorer people, and puts foodstuffs and cash income directly into the hands of the poor.

The efficiency argument goes back to the Nobel Prize winning work of Shultz, whose discussion of 'penny capitalism' argued that poor allocated resources carefully, i.e. their low incomes stemmed from a low asset base, not from failure to use resources efficiently and maximise market opportunities.

An important overlay on this argument was provided by, for example, Griffin, Lipton, and others who argued that policy distortions made items such as machinery and agricultural finance artificially cheap, lowering costs for relatively capital intensive commercial farms compared to smallholder farms, and consequently suppressing and disguising the underlying efficiency of smallholders. The policy implications are that the full benefits of efficient smallholder farming would only be realised with the removal of these distortions.

This view underpinned a line of argument within the liberalisation debates of the later 1980s and the 1990s, much used by the World Bank: that much reduced government intervention in the economy as a whole, and even in some aspects of agriculture, would likely have positive results for smallholders. The equity argument is based in the obvious point that smallholders include many of the poor, and so direct gains by smallholders are equivalent to direct gains by the poor.

In addition, smallholder agriculture may have stronger linkage effects on other categories of the poor than commercial agriculture or, indeed, most other economic activities which might form the basis of growth. These linkage effects include:

(i) direct upstream and downstream linkages, i.e., the sale of inputs to farmers and the processing and marketing of outputs;

(ii) labour market linkages, as prospering smallholders are more likely to employ extra labour, rather than replace labour with machines;

(iii) consumption linkages, as farmers will spend a relatively high proportion of their extra income on locally provided products and services (e.g. building expansion and maintenance); and

(iv) investment linkages, as smallholders invest in such areas as children's schooling, housing, land improvements and tree planting.

A weakness in the 'win-win' argument is that it gave insufficient prominence to market failures and focused instead on policy distortions. Essentially, the huge strategic potential of smallholder agriculture was seen as being held back by: the factor already identified, e.g. government policy distortions which artificially lowered input prices to more capital intensive sectors, e.g., commercial agriculture and other non-agricultural activities, while the predominant labour intensive sector, smallholder agriculture, faced relatively high input costs and, sometimes, artificially depressed output prices.

The depression of output prices resulted from either state marketing monopolies and/or exchange rate overvaluation. Overvaluation turns the internal terms of trade against traded goods and services in favour of non-traded goods and services. Smallholders may suffer more than most sectors because their output is predominantly tradables or semi-tradables, but their high labour intensity means that they make little use of traded inputs.

Of course, it was also argued that government needed to support smallholders through technology research and extension. While for many years there has been an awareness of market failures in poor rural areas, fairly recent developments in new institutional economics have helped in describing their causes, and therefore have assisted in thinking about how 'structural' these may be, and what autonomous developments or government interventions may be necessary to remove them.

The immediate point here is that if market failures are in some sense structural or embedded, then this must challenge policies which assume that smallholder development will take off simply from the removal of policy distortions and marketing monopolies, plus improvement in the quality and focus of research and extension. What the present author and Andrew Dorward have described as the Washington Consensus on Agriculture (WCA) essentially accepts the 'win-win' view as the premise for policies.

To unpack the arguments about the structural nature of market failure requires a detailed understanding of the challenges of contracting (i.e., of doing business) developed at a general level by North and Williamson, and applied to poor rural areas by, for example, Stiglitz, Bardhan, Jaffee and Morton, Dorward, Kydd and Poulton.

The overall perspective is that transactions activities (doing business) should be given equal prominence in economic analysis as transformation activities (making or growing things), because without transactions, only very basic transformation can occur.

A fundamental weakness in the classical tradition in economics, and both its neoclassical derivative and Marxian partial-derivative, is that there is weak analysis of transactions. Neoclassical analysis has tended to ignore transactions costs, or to treat them simply as a form of production cost.

Marxian analysis has seen transaction activities as unproductive, and has principally been concerned with analysing these as a means to structuring power relationships and thus facilitating unequal exchange or surplus extraction. In contrast, new institutional economics stresses that contracting entails a number of specific problems and that institutions are organised partly as responses to these problems, and partly as a means to rent extraction where power relations are unequal.

The specific problems of contracting include: searching for possible counter-parties with appropriate characteristics ; screening counter-parties for the desired characteristics; measuring what is to be exchanged (in terms of quantity and qualities); structuring the deal to maximise the chances of satisfactory performance; monitoring performance of the contracting parties; and, finally, enforcing the contract where there has been alleged noncompliance. All of these challenges demand information in abundance, and the structuring and enforcement aspects work best when supported by effective tribunal processes, backed up by legal and/ or social sanctions.

A key cause of market failure, or very weak markets, in rural areas is that the costs of the information needed to make effective contracts are prohibitive in relation to the potential benefits of the contract. This consideration is exacerbated where contract enforcement mechanisms are missing, as extra reliance has to be made on the personal characteristics of the counter-parties, i.e., more time has to be put in to assessing their reliability if subsequently there are no reasonable prospects of taking them to court for failure to perform.

Thus in a weak legal environment, more complex contracts are typically based on extended personal contacts between the counterparties, and this much reduces the possibilities for rapid expansion of a particular transactions activity. Institutions are rules of the game which have been devised to govern transactions, and it is obvious that the lower the transactions costs then: (i) the less the likelihood of market failure; and (ii) the greater the benefit to the counter-parties of the transaction.

Thus a development-enhancing path of institutional development will be one of progressive institutional innovations which reduce transactions costs. However, a key point of North is that we cannot assume that the path of institutional development will be transactions cost reducing. Powerful interests may wish to restructure institutions with the objective of serving their own short term interests, which may be achieved by increasing transactions costs through such devices as monopolies, taxes and other restrictions on contracting.

A key insight of Williamson is that economic activity is organised either through markets or hierarchies. Market relations are based on contracts, in the general sense of the term. Hierarchical relationships are often sanctioned by contracts (e.g. contracts of employment) but work on a day to day basis through individuals being prepared to accept and act on 'reasonable' orders issued by their superiors.

Both forms of organisation have advantages and drawbacks. In market transactions, there is usually greater clarity about the exchange, but sometimes very high transactions costs in relation to the value created by the contract, and also considerable inflexibility, resulting from the requirement to specify exactly what is being exchanged.

Hierarchies are, in some ways, more flexible modes of organisation, as individuals within a hierarchy are supposed to get on with executing 'reasonable' commands, without such detailed specification of quantities, prices etc. On the negative side, there are major supervision and enforcement problems in hierarchies. While competition is often found within hierarchies as a tool for promoting hard work and innovation, this may be ultimately less effective in this regard than inter-firm competition.

In Williamson's view, in a liberal society where government does not impose a pattern the frontiers between market and hierarchy will fluctuate, affected by the pace and direction of institutional and technological innovations. Smallholder agriculture is constrained by its organisational form to be distinctly more market than hierarchy. Tendencies towards concentration, which would essentially mean the end of smallholder farming, are often held back by social institutions such as usufruct tenure, sometimes supplemented by formal legal provisions to discourage farm amalgamation. In the last half-century, different institutional models have been tried to wrap types of hierarchy around what remains essentially a mass of decentralised small units.

At present, with the exception of contract farming and perhaps independent farmer organisations, none of these approaches are sanctioned by WCA theory, which quite explicitly advocates an institutional model for smallholder agriculture based on minimal hierarchy, i.e., independent and competitive contracts between smallholders and numerous independent suppliers of finance, of inputs and purchasers of produce.

State involvement in provision of marketing services to smallholders is strongly disapproved of under the WCA. In those states which have required extended balance of payments support from the Washington Institutions moves towards dismantling state marketing have often been a loan condition. There are grounds for doubt whether the implicit WCA path for smallholder development is feasible in areas of poor smallholder farming.

Output markets may work tolerably, in the sense that crop surpluses will generally be purchased if placed for sale in a local market. But financial markets for inputs are very problematical: lending for agriculture involves accepting climatic, pest and price risks, all to borrowers who have little or no collateral. As noted earlier, lending based on personal relationships is an alternative to collateral, but it is expensive in lender time, with tight limitations of capacity to expand the number of borrowers per lender. With weak or absent markets to finance inputs, farmers have to rely on financing from non-farm enterprises, remittances, asset sales etc. The problem is greatest in areas of unimodal rainfall (unfortunately the environment of many of the world's poorer rural areas) where it is more difficult to develop off-season farm enterprises to generate cash to pay for inputs for the next season.

With weak financing of inputs, it may be expected that input markets will be anaemic, with low density of outlets, limited stocks, low competition and high prices. As noted earlier, contract farming for certain cash crops may provide a solution, working because the provider of finance is able to recover this through the price paid for purchasing the crops.

Furthermore, financing of a cash crop may indirectly allow for some inputs to be applied to food crops. However, the contract farming model depends on the presence of quite demanding conditions, in particular, an assumption that the supplier of inputs has sufficient power (e.g. through a local monopoly) to force the farmers to sell produce via themselves, in circumstances where opportunistic buyers may appear and offer a higher price.

## Implication for Research

It is widely accepted that private profit motivated agricultural technology companies are not strongly attracted to the development of technologies appropriate to, or inclusive of, smallholder farmers because they do not represent a major market, especially in non-Green Revolution poor countries. The problems include the poverty of smallholders, their semisubsistence orientation and agro-ecological heterogeneity.

For many years, solutions have been proposed which would use government subsidies to encourage technology companies to engage with smallholders, and a recent contribution in this area has been made by Kremer and Zwane which has stimulated significant donor activity. An institutional perspective identifies an additional gap in this area, which is that appropriate transactions cost reducing institutions need to evolve in tandem with technological development.

The key implication is that institutional development and technological development are both necessary for smallholder agriculture to play its strategic role in poverty reduction. The larger scale private sector has a very important role, as it has in technology development, and subsidies should be offered for a package of institutional and technological developments.

Kremer and Zwane's work, and the policy and institutional development implications that follow from it, are narrowly focused on one market failure, that for technology development. Yet market failure issues surrounding smallholders are much broader than this. In other words, while it would be a great achievement to bring in policy and institutional changes which would stimulate the development of technology appropriate to poor farmers, the potential of this achievement will not be well exploited unless other market failures, e.g. in agricultural finance, are addressed.

The conclusion is that we should have a much broader view of how to act against market failure. While it makes sense to temporarily subsidise the private sector to overcome market failure, subsidies should be for effective delivery of components of the larger package of institutional developments that are required.There is much dissatisfaction with the effectiveness of public sector agricultural research in poorer countries, notably within the non-Green Revolution, nonglobalising group which contain 2 billion people.

There are voices questioning whether governments and development agencies should give up on public sector agricultural research, this thinking being motivated by views that: (a) the strategic role of agriculture in poverty reduction has been greatly exaggerated and therefore it is preferable to use research funds to back mainly non-agriculture-based livelihoods as routes out of poverty; and/or (b) 'globalisation and liberalisation will ride to the rescue', i.e., by connecting technology companies to small farmers, the former will start focusing on the latter as an important market segment.

In contrast to these views, it is argued here that smallholder development is strategically important to the earlier stages of successful poverty reduction, especially for the poorer non-globalising countries, yet globalisation may hold limited prospects as an avenue for development, due to pervasive market failures, and unhelpful evolutions in rich country consumer demands and supply chains.

## Economic Reforms and Globalisation

The rural economy is still the mainstay of India and it cannot be ignored. If the problem of poverty is to be tackled the next generation of reforms cannot afford to overlook rural development. In the urban sector the large metropolitan cities are the most immediately affected on liberalisation and globalisation, with significant changes in the land use and work patterns.

The claims made during the beginning of reforms that it is going to bring about employment growth does not seem to be true after studying the data for the post reform period. With more than 90 per cent working in the unorganised sector and the employment opportunities declining there seem to be urgent requirement in the policy changes to revitalise employment generation in the economy.

## Rural Poverty in India

The central and state governments are faced with severe financial constraints that restrict them from incurring expenditure on infrastructure building. The economic reforms had opened up many state-monopolised sectors for private investment. But the rural infrastructural sector gets very marginal attention from private investors. The lack of attention from the state and from the private sector may be deteriorating the rural infrastructural facility.

The growth of the rural economy is intricately related to the infrastructure availability, which strengthens modern economic institutions that are vital to employment generation and income growth. Evidences from studies show that the likelihood of a household being poor is influenced by the social and economic infrastructure of the village, agricultural technology used and demographic and other characteristics of the household. The Tenth Plan Approach Paper identifies rural infrastructure as a priority area, which requires urgent attention.

### *Urban poverty in India*

The opening up of the economy has seen the resurgence of the importance of large Metropolitan cities. Private investment, both foreign and Indian have tend to be concentrated in and around these large cities. The local governments are wooing these investments by doling out a range of incentives. The large metropolitan cities are undergoing a face-lift exercise as part of the city cleaning, beautification and pollution control programmes. While the city spaces are being increasingly acquired by the private commercial and service industry establishments the poorest, mainly the slum dwellers, hawkers, destitutes, street dwellers are being pushed out of the city to the peripheries. The city peripheries are getting degenerated with low value employment, poor living condition, thus making the lot of the urban poor worse.

**Employment and Poverty**

Opening-up in the developing economies was primarily visualised as a mechanism where trade would function as 'an engine of growth' and the fruits of growth would 'trickle down' to the poor. However, the results had been mixed, with many countries observing widening inequality in their economies, contrary to the conventional trade theory prescriptions.

The internationalisation of trade has opened up vistas for Globalisation of production creating profound changes in the labour market, such as widening wage disparity, increasing contractualisation of work, skill based segregation of work etc. In this context it is essential to understand the impact of opening up of the economy to trade and Globalisation of production under the new international economic order on the working poor in a developing economy like India.

The unorganised sector is called so because the activities in the sector cannot be accounted statistically. And they are generally accounted as a residual of the organised sector. But this residual sector is more than 90 per cent of the total work force in India as per 1991 census. This exclusion from the numbers have reflected on the policy level too, with hardly any legal backing, any social spending, or any other form of support to this class of workers, who are also the poorest among all groups of workers. Even among the workers there are no collective bargaining institutions to project their case. For the vast majority of them there is no fixed place of work.

no fixed working hours, no regular wages, no job security. Thus they have become one of the most vulnerable to poverty. Globalisation is argued to be 'informalising' and 'casualising' the employment opportunities in the economy thus further expanding the unorganised form to employment.

Women form a substantial and increasing share of employment in the unorganised sector. The world of work for women is characterised by unequal reward for equal work and uneven opportunities for the participation in economic decision-making. Added to this is a near stagnation in the number of female workers as a whole and an absolute decline in the number of female workers in the rural areas due to, declining work participation rates in both rural and urban areas in the post reform period.

Recognising that unorganised sector is an essential feature of developing economies and reconciling the fact that economic reforms are only going to vitiate the status of the sector it is to be understood as to how certain amount of economic security can be brought into this sector which would ensure them escape from poverty. It is also essential to understand the underlying factors that have vitiated the condition of women in employment and consequently poverty.

# 6

# Social Mobilisation and Women Empowerment

The recognition of women's powerlessness and addressing it with the base of low economic status of women is internalized at all levels in the organization. During different deliberation with Dhan functionaries and even during discussion with women members clear message of understanding on women's issues was observed.

Basic issues to be addressed are internalized in the system that women are restricted to low paid work, lack of access to education, training and credit, lot of invisible and unpaid work, restricted mobility of women, lack of any opportunities to express women's abilities, gender discrimination intermixed with cultural issues. With this understanding, community-banking programme for women has moved with clear reflection of fundamental core values and purpose of Dhan. These core values and objectives are reflected in each of Dhan's initiatives and actions.

Discussion and the review of the documents indicate that Dhan has actually not used the word 'empowerment' though it does realize that these are empowering processes happening. The outcome of its processes is seen more in Access and Control and decision-making through democratic processes. The strong belief is to allow the women to learn in their own way and not short-circuiting the processes. The ultimate goal is selfreliance and empowerment through interdependence with community. For this purpose, Dhan has adopted four generation strategy achieving one after the other.

The first generation process is social intermediation followed by financial intermediation process in second generation with livelihood and business promotion during third generation and finally the fourth generation process will address civic programme interventions. Dhan itself has perceived the time frame of 10 to 12 years for four generation but to the researcher's mind certain process were evolving due to women being together and women are demanding to address the civic issues like health intervention. The generation gap is thus reducing over the years.

Therefore in this sense, empowerment may be interpreted for Dhan as the process by which the women and community gain control over social and economic conditions; over democratic participation in their communities and over their own stories. Dhan thus believes that social change will evolve over a period of time with economic strengthening and thus views economic empowerment as the beginning of the social change. Dhan hence has adopted the economic entry approach.

Dhan foundation is a spin off institution of PRADAN, one of the foremost development agencies in India. Asserting a clear mission, working on value based principles and fundamental and associating professional staff with high values and commitment. An organization promoting 'participation of poor women' of course has in-built system of including participation of its programme staff. Dhan recruits professionals who are willing to work for the development. Various consultative processes existing indicate that all the staff are also allowed to reflect their voices and concerns through structured and unstructured interactions.

These interactions are further developed, refined and also validated through their experiences of working with the poor people. Programme staff is considered a critical link to work with the women. The strong qualities of perhaps each programme staff is a scientific method of social analysis, continuous learning, two way communication skills and the ability to cope with tension and conflict resolution skill mixed with strong sense of ownership and fundamental values of Dhan. However, the researcher felt that the long hours the team puts in on a day to day basis should be having pressures and stresses of varied kinds. The noteworthy point is that Dhan has a policy of gender balanced recruitment unlike other organizations who appoint only women for women's programme. The thrust is on recruiting and grooming the professional suitable for the development sector.

## Community Banking Programme

The focus of community banking programme is to create sound local financial institutions managed by women at each hamlet, cluster, & block level to link with main stream financial institutions.

### Programme Components

- Organizing poor women into groups of 15-20 and giving them an opportunity to experience and share their economic position and the nature and causes of exploitation - Social Mobilization;
- Mobilizing of own resources and managing them for their own benefit through regular savings and management of credit;

- Regular interactions, exchanges and exposures;
- Linking the women groups with formal financial institutions; The credit management is by women groups and the transaction with bank as clients and not as beneficiaries;
- Training and building leadership qualities through transferring the roles and responsibilities to manage their own groups and activities;
- Skill building and training in specialized business activities (Dairy, Poultry, Charcoal, etc,) and financial management;
- Formation of women managed financial institutions at village, cluster and block level for gaining collective strength and Power.

**Structure of People Institutions promoted by Kalanjiam Programme**

Community banking operates at group level, at cluster level and at federation level. Each group has three leaders decided by the consensus of all the members. The executive committee thus manages the group, which consists of the president, secretary and treasurer, a vice president and joint secretary. The secretary will be responsible for looking after operational and administrative matter and the treasurer would take care of cash and accounts. The executive committee would divide the following operational responsibilities among themselves viz. Banking, community issue, monitoring and evolution, expansion and training.

Once a set of 5-6 groups in a contiguous area has attained stability in their operations, they are encouraged to meet periodically and share their experiences. Over time these meetings are formalized and a cluster association is formed. The task of cluster association is to monitor the progress of the groups, conflict resolution promoting new groups and also training group members.

This has further evolved towards helping groups in creating bank linkages and also a forum for addressing the civic issues. Each group has three representatives in cluster of which seven members executive committee is elected at cluster level. Cluster constitute of 15-20 groups. The federation is made of 150 - 200 Kalanjiams. Federations get support for three years from Dhan.

All the groups will promote in to federation at the block level. These federations are registered as society or trusts. The general body of the federation consists of all the leaders of the member groups. From which 11 members are selected for the constituting executing committee of federation board. The primary objective of the federation is to strengthen and sustain the primary groups and mobilize external funds from banks, and apex financial bodies. Federation acts as a support body for training and managing specialized services such as insurance, health, and education.

Federations provide economies of scale, better bargaining and negotiating power with banks, government, outside agencies, while facilitating local initiatives and promoting leadership development. Within the overall structure each level operates as an autonomous body. Each level has to cover the cost by itself.

The autonomy is to ensure the responsibility of the group and group leader, strong ownership, awareness of cost coverage, developing community & financial management skill and also to ensure transparent system. These are empowering processes, leading to devolution of qualities essential for development of the women. For e.g. leadership- quality asking questions, negotiations, analyzing and consciousness to monitor the groups money.

The present structure of groups also allows for the participation of women at all level. A norm of one person - one post is followed to avoid burden on one person. Normally SHG leaders graduate to cluster leader and cluster leaders graduate to federation leaders. This process brings in the experience of the local context and also ability to share more responsibility by the leaders. Once this graduation happens the earlier role is taken over by the new member.

This process assists in rotation of the leadership and gives opportunity to all the members to participate and share higher level responsibilities. While women members take governance roles, local staff are appointed for maintenance and management functions. The costs of the local staff are paid by the people out of their own sources. There are around 1500 local staff working at group, cluster and federation level.

At present a total of Rs.30 lakh per month towards the salaries of these staff are paid by people. In essence these groups are also creating employment opportunity at local level. One group accountant keeps the record for 5-6 Kalanjiams. One of the important things is to be noted here is that the money transaction does not take place through the hands of accountants. They are merely facilitators for writing the accounts and record keeping. Similarly the cluster appoints cluster accountants and cluster associates. The women members primarily are landless agriculture labors, masonry workers, and construction workers or undertaking small activities of fruit and vegetable vending or running a petty shop.

The classification of families is categorized in survival, subsistence and self-employment. Almost 80% of the women belong to survival and subsistence category having wage income with deficit household budget leading to greater indebtedness. Village mapping and wealth ranking in the village under take identification of the women members. The Kalanjiam is a basic hamlet group with 15-20 members residing in the same hamlet. The programme is implemented by mobilizing the target women in homogeneous group.

Homogeneous group here means organizing group of target women with more or less similar economic status. The first generation group formation took longer time sometimes 5-6 months to formalize the group. Dhan staff made the initial facilitation efforts through addressing the social issues. In KKVS federation issues like girl infanticide or suicide were addressed to build confidence and rapport among the women. The second generation Kalanjiam formation was motivated on the experiences and example setting of older groups.

Over the years, this process has shifted to the group members and the group leaders. Many Kalanjiam have emerged as a ripple effect of the groups functioning in that area. This has now transformed into institution building through Kalanjiam movement exclusively constituting of poor women. Kalanjiam movement of group expansion has emerged in to creation of women led managed kalanjiam Foundation.

The process of initial grounding goes on for 3-4 months and follows process of "quality check" followed by the registration process (not legal body) at programme level for allotment of group code. Books of accounts are introduced after six-month period with opening the bank account. The norms for group functioning such as saving rate, loan amount, interest rate, meeting schedule, agenda setting and group leadership are decided by the group.

The members as convenient to them decide the timings of group meetings. In semi urban areas the women prefer to meet generally at 10.00 am in the morning as they are free of household work. Most rural groups prefer to meet in the afternoon after becoming free from the work. Frequency of the meeting is monthly or fortnightly, usually monthly in rural areas. Fisher women's group in Ramnad district meets four times a month.

Meeting place is usually outside the village under the big tree. In semiurban area women meet in a common garden or common open space of the hamlet. One of the groups is meeting at the temple place, which also happens to be the house of village panchayat leader (Man).

Due to the temple place, women during "periods" are not attending the meeting. In one more group few women do not attend the meeting for the same reason as this is held at one of the better of members' house. At this juncture women should negotiate to obtain space for "Mahila Kutir" to have common place of meeting which may also serve the purpose of shelter for the women.

The attendance in the meeting appears regular and member not attending the meeting will take permission or inform. Some groups have the practice of imposing penalty for not attending the meeting. It was observed that the participation by the women in the semi-urban area or town areas was higher. The participation also depended on the literacy level. Normally, the younger women in the age group of 25-32 years are more participative, initiating the issues and also playing the facilitator's role.

The older women are matured and able to understand the issues. Extremely poor category, sickly or very many elderly and totally illiterate women are seen excluding themselves from the discussion. The women in semi-urban areas are seen more articulated and organised in conducting meetings and sitting in a round sitting arrangement. In the rural areas even now, articulation is less and does not tend to sit in a way that allows eye to eye contact.

Erstwhile Pradan project initiated 1989 made a modest beginning and started with forming 20 self help groups in 11 villages with a total membership of 286 women. This has under gone various phases of expansion with different strategies like initial promotion and strengthening, older group mobilizing new groups, replication in new location and saturation in a particular village and also in the block, revival of defunct groups and collaborating with other NGO. In the year 1998 a revolutionary movement started with a decade of community banking experience resulting in the beginning of people led movement.

The year witnessed the celebration of Kalanjiam movement and women taking oath to spread the concept of the Kalanjiam to all poor women and also to reach one Lakh women within three years. During the year 1999-2000 itself the Kalanjiam covered additional 33,000 families. The VVK Federation has made 100% coverage of all the poor in 39 out of 80 villages saturating the coverage of poor in the region. The effort is targeting all the poor in the panchayat and in a block so that 100% coverage of poor is ensured.

The expansion strategy is further strengthened by the initiative of cultural team, by movement workers and collaborating with local government and banks. This approach has enabled to transform from a mobilization and service delivery approach to people led movement and also cost efficient approach. The next strategy is to have a Kalanjiam Foundation a separate thematic institution to upscale the Community Banking Programme and reach a million poor over next decade. The major approach of the programme in meeting its objectives is through the promotion of people's organization at different levels.

The programme has been successful in promotion of more than 6100 SHGs in almost 2000 villages with 400 cluster development associations and 20 federations. As on December 2001 there are more than one-lakh Kalanjiam members. As a support body facilitating the development of local institutions and enabling the women to manage and become self reliant for own development, Dhan appears to be the only institution to achieve such a large scale.

The growth in terms of savings and lending activity is equally impressive. Most distinguishing is the fact that this is all poor people's money. Average saving is Rs 1300 to Rs.1500 while average loans arrive at Rs 2500 to Rs 2800 at their doorstep. Though the average saving and average loan will not the capture the strength of some of the older members but gives an indication of large scale outreach of the benefit of the services.

**Kadamalai Kalanjia Vattara Sangam (KKVS) Federation**

Kadamalai Kalanjia Vattara Sangam (KKVS) is located in Mayiladumparai, a remote block in Theni district of Tamil Nadu, 100 k.m to the North West of Madurai, in the foothills of Western Ghats. The block has 18 panchayats that have 115 villages among them. The main reason for selection for operating community banking programme was the remoteness of the block and its constituent village and low development status both social and economic. This development status is greatly influenced by migrant nature of population and the area being pre-dominantly dry agriculture.

The villages were selected on 4 major criteria remoteness of villages, small villages with less than 100 households, villages with more women issues and finally the prospectus of forming clusters due to contiguity of the villages. The federation was registered in August 1995 under the Society Registration Act 1860 after a period of 2 years of association with Kalanjiams.In terms of resources land and forest are the major resources of the block but with the degrading quality.

There are 924 females per 1000 males as of 1991 census indicating poor female male ratio. Scheduled Cast Scheduled Tribe constitutes nearly 24% of the overall population in the block. Of the total work force 46% are female workers while female agricultural labourers are 49% of total agriculture labourers. The literacy rate is very low at 35% with female literacy at 24%.The major development issues of the block are as female infanticide, the marriageable age of girls between 12-16 years, desertion of women by men, large number of widows, high incidence of money lending.

Some of the infrastructure issues of the blocks are high deforestation and high erosion, land alienation, migration of men in search of employment and extensive cultivation of cannabis. Similarly, lack of approach roads to villages, poor health and educational facility are other features of the block.

**Mugavai Kalanjiam Mahalir Vattara Sangam (MKMVS) Federation**

MKMVS is operating in the entire Mandapam block having 28 panchayat and two-town panchayat and in 7 panchayat of Tirupulani block of Ramnad District. The community-banking programme was initiated in 1992 in Mandapam Block and in 1995 the Kalanjiams got promoted into Federation. This block was selected for development intervention, as it is one of the most backward blocks in Ramnad. Livelihood of most people is dependent on coastal and marine resources. Fishing and palm mat weaving are major source of income to meet their basic and social needs. Agriculture is not suited for this block due to sandy soil. Special feature of this block is that the male population is less than female population. The population of Mandapam has religion based diversity as equal number of Hindus and Muslims are residing in this block.

Many poor Muslim males are migrating to Arabian countries for earning but most of the time they are cheated by the middleman. The socio-economic situation of the block is very poor. Different occupations have different wage rates in the same range of Rs.20 for women and Rs.35-70 for men. This block is also characterized by large number of women headed families. Alcoholism is wide-spread and men are spending their money only on alcohol and give meager amount for family expenditure at home.

There is a large scale desertion of women and prevalent bigamy. The villages covered under the block are remote villages not connected by pucca roads; inadequate drinking water and electric facility, exploited money lending. The general profile of the members covered are landless labourers, widows, prevalent child labour, female headed families, people living in huts and earning less than Rs.1200/- per family per month.

**Pothigai Vattara Kalanjiam (PVK) Federation**

PVK federation is located in Madurai west block having urban influence as it is situated around 8 kilometers away from heart of Madurai City, with community banking programme initiated in the year 1999. Nearly 60% of the total geographical area is used for agriculture purpose while rest is wasteland and used for non agriculture purpose like institutions and industries. The female population is 974 per 1000 of male population. Female agriculture labour force constitutes 45% of the total agriculture labour. The total female workers are 34% of the total labour force.

The female literacy is 44%. The main employment opportunity for the block is agricultural work. This block is one of the paddy belts in Madurai district. The wage rates vary from Rs.25-30 for women and Rs.50-60 for men. Main reasons for selecting this block for community banking programme is that 60% of people are agriculture labour, money lenders operations at exorbitant rate, high percentage of scheduled caste population and no infrastructure facilities.

Most of the people are employed as agriculture labourers, road and building construction workers, fruit and vegetable venders, mill worker or undertaking activities like coir making, dairy, and miscellaneous small business activities. The targeting of the villages was done on the basis of remoteness and population less than 1500 in a village of which majority villages targeted are with population of less than 500. Only two villages have population above 1500. This federation has attempted classification of poor into three categories of survival (who find it difficult to fulfill their daily needs), subsistence (wage employment available more or less regularly) and self-employed. The major development issues in this block are alcoholism, child labour, high dropout rates of children from the school and infanticide.

## Growth of Financial Services

Providing diversified financial services is considered a vehicle to address poverty. Widening of financial services is addressed by penetrating the market by geographical saturation through replacing the informal sources of credit. Dhan strongly believes that deepening of financial services is possible through providing options and choices for the poor. Once poor are utilizing the services it will help them to link and influence the mainstream system. Alternate financial services thus are provided in an organized manner with an understanding of varied dimensions varying in each local context and situation. Savings, credit and insurance services are provided to suit the convenience of the poor women.

Saving is considered a powerful tool to build the stakes of local community in the financial system. Started with level of Rs.5/- for each member, women are saving up to Rs.100/-. There was no concept of saving earlier and whatever was earned, was spent. The saving collection in the group is weekly, fortnightly or monthly. In Mandapam block in Ramnad District groups have achieved the saving level of Rs.10 million. The total savings of all the Kalanjiams are Rs.1067 lakhs as of March 2001 as compared to Rs.627 lakhs in March 2000.

The average savings by member per year increased from Rs.55 in 1991 to Rs.337 in 1995-96 and has risen to Rs.1329 as of March 2000 and to more than Rs 1500 by December 2001. In older groups the average savings range from Rs.3000/- to Rs.5000/-. The saving lending ratio for all the Kalanjiams comes to 51%indicating the extent of people's own contribution for rotating the funds.

Two types of saving products are available primary saving and diversified savings. Primary saving is the regular contribution and internal lending is used for rotation within the group with out any lock-in period. Primary savings are non-withdrawable until the death of the member or on discontinuation of the membership from the group. The withdrawble diversified savings are contributed for specific needs like education, marriage, festivals, social obligations etc.

There is a special deposit scheme allowing withdrawal (twice the amount) after a fixed period. The diversified savings take care of seasonalities and ensure food security during emergencies. The flexibility of differential savings enables the poor women to save according to their needs and also on the basis of ability to earn. Many groups during the lean season save differential amount.

Many microfinance organizations do not emphasize on savings. While for DHAN Foundation savings emerges as wherewithal for viability of groups.

Community banking programme provides range of services to meet different credit needs of the poor including consumption, health, education, income generation, housing and asset creation. Credit for food security and for social security like health education is considered credit for consumption. Usually, during first year, the loan is given eight times of savings while after first year five times of savings is considered for credit. The number and amount of loan of each member increases as the age of group matures.

There is no tangible collateral requirement for availing loan for the women. Indirectly Savings works as collateral as it gets adjusted incase of default of loan. Installment repayments are fixed flexibly on the basis of purpose and quantum of loan. the older federations the individual loan amount has increased to Rs.20000 - Rs.25000. The bigger loans are given for house repairs or construction of house.

In these cases it is ensured that asset is created in the name of women only. It will be advisable if these members are encouraged to open individual bank account to avoid handling large cash on one hand and on other to inculcate the habit of direct mainstreaming with the bank. The major impact is noticed in redeeming outside debt that is availed at the rate of 10-15% per month.

Moneylenders refuse loan to very poor unless they pawn their small assets and that also under great humiliation. However, a few women have stated to be saying that flexibility of very small amount like 50 Rs for very short period is available only with moneylenders.

## Social Mobilization and Empowerment Impact of Micro Finance

The sharp visible impact of the Community Banking Programme is seen in its outreach to the poorest of the poor in backward and remote locations and curbing the moneylenders' activities of up front usurious interest rate. Substantial resource mobilization and more than 98 % repayment pattern with negligible non-performing assets are other quantitative indicators helping to further upscale the programme.

Qualitative indicators like cadre of poor women as leaders, strength of collectives as social support structure creating an identity of women in the home as well as in the village, confidence, physical mobility, access to public space and increased control on financial resources are to be observed and believed. The biggest impact is seen in successfully creating Kalanjiam movement led by the poor women and has given the poor women a sense of achievement. The programme has provided the space and opportunities for the poor women.

The saving by poor women facilitated greater control over financial resources and access to cash in need. This has enabled the women feel more secured during emergencies. Women take pride in owning of assets in their name like land and house through loans. Access to productive assets like land lease and tree lease through credit utilization is available to women for undertaking doing business activities.

Women do take the consent of men for obtaining the loan and for the purpose for which it is to be utilized. Undertaking self employment activity like petty shops, land cultivation, dairy activity, Small trading activity, flower selling and rope making etc have been initiated by many women but are not seen on large scale.

Most of the time self-employment activities become the collective activity for men and women though there are examples of women taking up the activity solely There are also examples of women like Kartamma saying "I cannot afford to own any asset from loan as I need to spend for my daughter and grand daughter as my daughter is widow at 20 and son in law committed suicide he was in love with other woman." It is also noticed that wherever women are doing independent activity, they have better control in decision making as compared to women doing labor work. Perhaps this is so due to differences in wages of the men and women.

Due to access to resources women's status has enhanced to a small degree and the bargaining position in household has become little better. Repayment of installment becomes joint responsibility. Women have discussed that there are tensions and conflict during tough time and at the first instance men will like to curb on the repayment of loan installment. A Kalanjiam cluster leader stated, "Men allow attending meeting in kalanjiam as suddenly we have become source of money".

A feeling of resistance among male for attending the meetings previously has changed. Wherever the male opposition still persists it is noticed that women are ignoring and attending the meeting. Men now support and wait at bus stand with bicycle for the women coming late after attending the meeting. Graduation of leadership has enabled the women to share larger responsibility and decision making. Many cluster and federation leaders have to spend lot of time for meetings, monitoring, group formation and other managerial work are happy to forgo the wages for the day they are giving for this programme.

Women headed household due to (widowhood, desertions, single women) are perforce taking their own decisions. They are taking greater leadership roles and have greater degree of control over assets and income. It is noted that in such cases when the son grows up he starts controlling on mobility and income of women and tend to load with household work also increases. This indicates gender subordination passing from father to son. Position of women headed households due to male migration is more or less same.

Poor generally eat rice, a regular staple food in this area. Therefore decision making in food preparation does not have much choice. In Mandapam Block women did mention that men demand cooking fish every day while children and women are happy eating rice every day. Women said "We have to cook what they want". Decision to save is influenced due to availability of repeat credit. Rural women are able to contribute from their wages. Women in semi urban areas save by curtailing certain expenses like cutting down on Tea or cutting on tabbaco chewing. Sometimes men help in contribution of savings but they do not seem to be cutting their unnecessary expenses.

Intra household change analysis indicates that in semi urban area women are able to speak and assert their rights. They have become bold and much articulated. Women are able to put up their demand forcefully. Kalanjiam has enabled them to get social and emotional space to shed their tears. Most poor in rural area cook rice once a day in the evening and manage till next after noon. In such a situation it is difficult to comment on nutritional status & also difficult for them to discriminate between boy and a girl.

Access to private resources has given sense of pride and ownership along with increased selfconfidence.

Above changes are more reflective in leaders and in those who have availed different cycles of loans.

**Educational Opportunities**

With the introduction of Kalanjiam Community Banking Programme an effort is made to introduce functional literacy. 70% of the Kalanjiam members are able to write their names and signature while 30% are still using thumb impression. Women do feel the need and importance of education for girls but do not make sustained efforts for girl's education. A few cases are seen at town/SU center that girls are studying up to 12th standard. Education loans are taken for the higher education of son but not for daughter though she is doing well in her studies.

A feeling still exists that "It is waste to educate the girl as she will be married." The general view is that by giving education the person neither remains fit for village life nor for city life. Probably identification of new livelihood sources with vocational or technical training may be considered. In all the areas visited, women are aware of child care services and do send the children to ICDS centers and also avail the facility of supplemental nutrition for lactating mothers.

However women are not aware of the role of these centers. At few places women did complain about the quality of food but never seem to have challenged. A close convergence between the programme and ICDS centers may be advisable for obtaining better services of ICDS. Women do support the PHCs during polio vaccination campaign but do not seem to be utilizing or challenging the services. First aid training to 59 women in VVK Federation has enhanced the health & hygiene awareness and these first aiders are able to serve 59 villages in their area. TBA (Traditional Birth Attendant) training has helped the women to conduct safe maternity but these developments are area specific.

Access to training on administration and financial management has given the women a sense of sharing responsibility and ownership. Individually women are more aware, have got an ability to analyze, questioning and negotiating, cost conscious and moreover enhanced confidence to deal with tough situation in day to day life. Co-learning and sharing experiences in Kalanjiam gives them ability to understand each other.

In urban federation women have accessed to new technology of accessing low cost housing. Women have started building a training center with the support of HUDCO. In rural federation access to new technology is not availed.

In MKVS federation (older) women did mention that we now need skill building training to earn more. Providing vocational training or BDS support is not presently done by DHAN. Women are able to access bank credit and talk boldly with bank officials. With the good repayment record these women are promptly attended. Women feel that over the years attitudinal change is observed towards the poor. Women are telephoning the bank manager demanding the status of their applications.

Apex bank linkages through federation are accessed for asset building and for income generation activities. However women have taken a note, "apex banks do not recognize us individually and are giving loans due to Kalanjiam and Federation". Access to housing loan and apex bank linkage has enabled number of women to own houses in their own name. This facility is available but many women need to utilize this facility.

## Labour and Income

In order to understand the nature of waged and non-waged work and also the workload, an exercise of mapping of daily routine was undertaken for men and women in all the three federations visited. Shockingly, despite working for 14 hours a day, women did not admit that they were working more than the men. Most of the men are found not working for more than 7 hours but considerd harder working and bread earner for the home. Income of Women on the other hand is considered supplemental. Non wage work is a household work and there is no autonomy to choose in this unless girl child starts helping the women. No choice is available to women in seasonal agriculture labour work as women only do weeding and transplanting.

The men do object to non wage work of women for instance taking the responsibilities in Kalanjiam and spending time for that. Looking at the benefits men have stopped reacting. According to husband of one of the cluster leaders "my wife is already overworking at home. She should not take additional responsibility still objects her owning responsibility."

According to the same cluster leader her husband says, "it is our money and we are getting loan on therefore there is no need to spend so much time for Kalnjiam". This actually reflects the mail ego and jealousy due to that fact that she is much-respected leader and getting recognition. These cases may be sporadic but sensitizing the men on entire issue of the programme may be considered through Kalanjiam foundation. Time spent on different activities

Mapping of daily routine also revealed that as soon as man is free he would take bicycle and go to teashop or in a common place where men assemble. For women eisure is to sleep and now watching TV and nothing beyond that. Now at least women are able to spend time in Kalanjiam, which is generally two times in a month. Women doing self employment activity feel they are putting in more efforts to see that it does not make loss.

Men's participation in household work is found in few cases but this seems to be more related to the gender and age composition of the children. Men do help in fetching water and breaking the fuel wood. This remains clearly within home and women were quite reluctant to talk about it. In majority of the cases the men do not share household work.

Wages are unequal for men and women as women feel that men's work require more strength. Wages for men are Rs.40 toRs100 while for women Rs. 25. Men and women are able to work for six months as agriculture laborers. Men try to look for alternate work like construction work or migrate to but women have no work during this period. Therefore women think men's wage contribution is more important.

## Socio-economic Environmental Changes

A change in approach towards women is distinctly visible at all levels. A positive change in terms of attitude and practices of mainstream financial institutions, government departments, weakening cast and religious barriers is notably conspicuous. There has been substantial reduction in turn over of local moneylenders. In fact this is one of the major achievements of community banking programme. There have been instances where the moneylenders have been challenged by the Kalanjiam members for misappropriation of poor people's money or exorbitant charging of interest rate (10% a month). Money lenders did try to dissuade the women initially but in the process lost their BUSINESS.

Successive smaller loans have enabled the women to come out of debt and ensured them the security of food. At least they are out of negative cash flows. There has been protection of the existing income level but increase in income is felt only when women have taken the loans for business activity. There has been no perceptible change in the consumption pattern of the poor people. Women are aware about obtaining food grains from PDS. Evidences in Mandapam block are seen challenging the PDS shops for lower weights and long uneven queues but the local government ignores this.

The processes of impacting poverty are happening in individual cases and will take time to see drastic changes in reducing poverty profile. On the basis of observation it is noticed that nutrition status of the children appear satisfactory. There is definitely scope for groups to create systematic links with ICDS programme to strengthen health and nutrition impact. But there is general lack of awareness among the women on health aspect as information on reproductive health system is highly desirable.

The impact is also difficult to ascertain in terms of increased participation of girls in formal education. As referred earlier, the women are recognizing the importance of educating young girls. It is a point of concern as girls stop going to school as soon as they attain puberty. Girls are also dropping out of school to take care of younger siblings incase mother is doing self-employment activity. They do get back to school after a break but by this girls have lost the interest in studies. Boys loose interest after 8th or 9th standard and start getting in to labor work.

In semi urban and small town the trend seem to be reversing and illiterate women are keen to send the girls and boys to school. Child labor is the impact of the programme can not be said. On the contrary women take loan to help the son to study. Thus it can well be inferred that in terms of Kalanjiam an alternative women friendly financial institutions has emerged giving space to market leaders from village to district level. The biggest achievement of Dhan is in providing financial services to the poorest women. This however, has yet to make a considerable impact on poverty profile of villages; nonetheless the economic entry approach has made the beginning arresting the income leakages.

## Rural Micro-finance in India

Credit is sought for basic requirements such as food, as well as for income generation activities. The rationale of micro-finance is based on the hypothesis that the poor can be relied upon to return the money that they borrow. Moreover, the repayment will also be on time. It has been proved that the poor are capable of thrift and savings. It is these existing requirements and conditions that are tapped by micro-finance initiatives.

Micro-finance as a development initiative has been justified on the grounds that it is beneficial to both micro-finance institutions as well as clients. Since the poor can be banked upon to return loans on time, it is believed that micro-finance and profits are not antithetical to each other. In fact this intervention is being hailed as the one method that will address both development and promote market behavior. On the part of the clients, it is believed that it is possible for micro-finance institutions to instill trust in the poor to give up their savings and to avail of micro-credit so that they can pull themselves out of the poverty trap.

The need of the poor for credit is not new. So far, this need has been met largely by informal sources such as moneylenders, support by kith and kin, friends, employers and landlords. Borrowing from these informal sources often involves exploitative rates of interest and results in strengthening of systems of oppression. The formal sources of credit in India have been banks and poverty alleviation programs promoted by the government.

The track record of these formal sources has not been positive. Micro- finance, in the form in which it is being promoted currently, circumvents the drawbacks of both the formal and the informal systems of credit delivery. Among the real and potential clients of micro-finance, women are seen as the most reliable in terms of repayment and utilisation of loans.

The gender dimension of micro-finance is based on the understanding that the entire household benefits when the loans are given to women. Further, it is argued that micro-finance can empower women since it instills a perception of strength and confidence when the poverty trap is broken.

Most micro-credit initiatives require the formation of small 'selfhelp groups' (SHGs) of 10-20 persons, who come together with the intention of saving and rotating loans amongst the members. Once these groups stabilize, they are accorded formal support so as to widen their lending capacities. The entire process of forming a group, of functioning in a sustained manner, of regulating finances, and being mutually accountable, is in itself, empowering.

An important dimension of SHGs is the peer pressure, which the members of a group exert amongst themselves, which acts as a substitute for formal collateral in that it is taken as the guarantee for loan repayment. The initiative of providing credit to the poor is not a new one in India. In the mid-sixties, the Government set up cooperative societies to meet the credit needs of people. Unfortunately, for a variety of reasons, this initiative did not meet the purposes that it was supposed to fulfill.

Following this, 1969 saw the nationalisation of banks in the hope that the credit requirements of the poor could be addressed in a more meaningful manner. These efforts were supplemented with poverty alleviation programmes like the Integrated Rural Development Programme (IRDP), Jawahar Rozgar Yojana (JRY) and Pradhan Mantri Rozgar Yojana (PMRY), the results of which have also proved unsatisfactory.

Over the past three decades, women and their credit needs have been explicitly addressed by women's organisations like Self Employed Women's Association (SEWA) in Ahmedabad, Gujarat, Working Women's Forum (WWF) in Chennai, Tamil Nadu, and Annapurna Mahila Mandal in Mumbai, Maharashtra. Their efforts have resulted in far-reaching changes in the lives of their women clients. Even so, today, micro-finance is perceived as a paradigm shift in the quality of micro-finance delivery. Further, the push has come internationally, from donor agencies, UN organizations, and the World Bank, all of whom are advocating micro-finance as the banner under which development can be achieved and poverty can be eradicated worldwide.

Officially, the intervention of microfinance has been heralded world-wide as the best cure for poverty. There have been a series of meetings and summits that have been held over the last few years, to design an approach that can been be followed by all countries across the globe. The meetings worked towards contributing inputs for the World Micro-Credit Summit Campaign held in February 1997.

The South Indian Consultation was held in Hyderabad, India on, 23-24 August 1996. A meeting of 20 practitioner NGOs, development financial institutions and development professionals arrived at a common position and action plan. The Dhaka Declaration of the South Asian Coalition for the Micro-credit Summit articulates the collective consensus among 21 networks and agencies delivering financial services to over 4.5 million poor people across Bangladesh, India, Nepal and Pakistan.

It endorsed the importance of micro-credit and emphasised that microcredit should be approached as a socially responsible business. Prior to the Washington Summit, a group of NGOs and development finance institutions met in New Delhi on 23, January, 1997. The objective was to review the progress made and to discuss modifications to the draft documents released by the Summit Secretariat in November 1996.

## World Micro Credit Summit

The World Micro Credit Summit was held in Washington DC in February 1997. More than 2,900 people representing 1,500 institutions from 137 countries participated in the Summit. The Summit announced a global target of supporting 100 million of the world's poorest families, especially women, with micro credit for self-employment and other financial and business services by the year 2005. This Summit received impetus in the mid-1990s after the World Summit for Social Development in Copenhagen in March 1995. Four core themes were stressed as part of a 55- page Declaration and Plan of Action.

- *Reaching the poorest*: 1.2 billion people some 240 million families - in the world are living in absolute poverty. This is the group that is targeted by the Microcredit Summit. The Summit promotes the use of quality poverty measurements to identify the poorest.

— *Reaching and empowering women*: Since women are supposed to be good credit risks, and women-run enterprises yield benefits for their families, micro credit is seen as a tool to empower women.

— *Building financially self-sufficient institutions*: This theme is based on the experience of developing countries which have shown that micro credit programs can improve their efficiency and structure their interest rates and fees to eventually cover their operating and financial costs. The Campaign offered daylong courses at global and regional meetings held from 1999 through 2001, which trained practitioners in this regard.

— *Ensuring a positive, measurable impact on the lives of clients and their families*: Two impact evaluation studies conducted by an NGO, Freedom From Hunger, showed that current clients of its affiliate institutions in Honduras and Mali had experienced positive program impact at the individual, household and community levels. The studies showed the higher levels of empowerment of client households as compared to non-client households in terms of larger enterprises, increases in personal income and household food consumption, savings and a feeling of self-esteem.

These four core themes of the Summit Campaign help to focus on both the targeted poor households as well as on the quality of the practitioner's work. Papers prepared for the Summit, substantiated these themes and have since been updated and used widely as guidelines by practitioners in this sector. After the 1997 Washington Summit, by December 1999 more than 1600 micro credit institutions joined the Summit's Council of practitioners.

In doing so each institution had endorsed the Summit goals and agreed to submit an Action Plan within one year of joining the council. In November 1999, Institutional Action Plan grids were mailed for the year 2000 to the 1,600 institutions. In India the All India Women's Conference was designated to collect data for this survey. All practitioner-Action Plans submitted were reviewed by the Summit staff. In an effort to verify the data, the fifty largest institutions in Africa, Asia, and Latin America were asked to identify donor agencies, research organizations and networks that could take up this task.

## Institutional Mapping

### Availability of Data

The available data can be scrutinized using the vertical approach. In such case, one finds that the data emerges in six different dimensions. Data throws light on the following aspects of each of the participants in the institutional mapping:

- Status of the participant ( nature, composition, socio-economic profiles)
- Organizational working (democratic nature, nature and turnover of staff members, laws, bye-laws)
- Operational functions of the participant (availing of funds, disbursement of loans, mobilization of savings, saving- investment-credit products available, repayment rules).
- Performance of participant (financial health, sustainability, quantitative data about intermediaries, judgmental data regarding standardization, rating, regulation and monitoring of participants, especially MFIs )
- Operational dynamics (inter-participant, intra-participant, both in relation to conflict resolution in

the formation of groups and their relationships with other participants)

— Resultant best practice norms relating to 2-5

**Structure of the Micro-finance Sector in India**

The rural financial system in India has evolved through two sets of financial institutions - formal and informal. The formal system consists of a multiagency approach, comprising co-operatives, public sector commercial banks (CB) and regional rural banks (RRB), private sector banks. The public sector banks and RRBs constitute the government sector. The cooperative sector comprises of district cooperative banks (within which are primary societies and urban banks) and primary land development banks.

Private commercial banks constitute the third channel. Since existing private banks do not do much rural lending, the Reserve Bank of India has provided for the setting up of Local Area Banks in the private sector to promote savings in the rural sector and to provide credit locally. The rules and regulations within the financial sector govern these institutions. The informal system consists of rotating savings and credit associations (ROSCAS), traders, merchants, contractors, commission agents, moneylender's etc. Each of these are governed by norms and rules that have been formulated by the concerned instituion/person.

The major objective of the nationalization of banks in 1969 and 1980 was to improve the flow of institutional credit into rural households, especially the poor. Commercial Banks play a dominant role in rural financial markets in contemporary India. Priority sector lending was one of the most important tenets whereby the RBI stipulated commercial banks to ear mark 40 per cent of heir advances for priority sector lending.

Of this 18 per cent was for agriculture and 10 per cent for weaker sections. The rural credit market expanded significantly in the post- nationalization phase. There has been a significant improvement in outreach due to nationalization and social banking. Between mid 1970s and mid 1990s bank credit in the rural sector increased by nearly 240,000 million and rural deposits increased by about 410,000 million. Therefore, the share of both rural credit and deposits almost doubled during this period.

The poverty eradication programs of the government also played important roles in addressing the credit needs of the poor. The Integrated Rural Development Program (IRDP) was the largest ever anti-poverty program launched during the Sixth Plan. It used the banking channel to direct assistance, which was a combination of credit and subsidy to those below the poverty line.

However, trends also show that the accent on development banking of commercial banks weakened considerably in the 1980s and 1990s. There has been deterioration in the pace of increase in outreach of commercial agencies and a decline in non-institutional sources in the 1980s. There is also a deceleration in the growth of commercial bank credit to rural areas in terms of both disbursal and outstandings in the late 1980s and early 1990s.

Further there has been relative fall in the proportion of bank credit flowing to the priority sectors, especially agriculture, since the mid-1980s. There is a close connection between the commercial banks outreach to the rural economy and the government's commitment to anti-poverty programs. It may be argued that the so-called development outlook of the banking sector is in fact a reflected one and does not by itself signify its belief in the bankability of the poor.

The IRDP targets were achieved because the banks were compelled to do so. Several relaxations were made with respect to eligibility criteria, procedures, rate of interest, collateral security and guarantee for the loan etc. However, the viability of loans is open to question and the recovery performance has also not been satisfactory. A dilemma seems to exist between developing a sound banking system and simultaneously dealing with poverty.

There appears to be a mismatch between development goals and what is expected of economic/ financial prudence. Banks have generally shown disinclination to service the poor who are distanced spatially and metaphorically from them and hence whose risk profile is difficult to assess. In the recent years, economic factors eventually took precedence over the social agenda.

Regional Rural Banks which were set up with the primary objective of meeting the credit needs of the weaker sections, were advised to lend 60 per cent of their incremental credit to the target population in the early 1990s. This was later brought down to 40 per cent and by end 1990s it was 10 per cent. Further, frequent loan waivers and write-offs, which were steps taken in political interests, hastened the process of the erosion of the rural credit delivery system.

Now there seems to be a general agreement among the banking authorities and development professionals that development agencies in the non-governmental sector should be called upon to act as intermediaries in the delivery and management of rural credit. The formal sector consequently took the initiative to develop a supplementary credit delivery mechanism by encouraging non-governmental organizations (NGOs) to act as facilitators and intermediaries.

Currently, the rural micro-finance sector is dualistic in nature. The formal structure has a legal and regulated component, which provides credit and other services to the non-formal sector. The non-formal structure largely comprising NGOs, SHGs, clusters and federations of groups operate outside the legalized structure and have demonstrated considerable organizational flexibility and dynamism in responding to the demands at the grass roots.

The institutional mapping of the participants involved in micro- finance interventions spans from regulatory bodies (like the Reserve Bank of India and NABARD), to apex bodies that disburse bulk credit for onlending to intermediaries that are involved in providing financial services to clients, and finally to the clients themselves who are the beneficiaries of the initiative.

## Institutional Mapping

Within the institutional mapping of micro- finance participants in India, the players constituting the regulatory bodies, apex bodies, and intermediaries, comprise those responsible for the supply side of the intervention. The clients constitute the demand side of the intervention. The regulatory body is primarily the Reserve Bank of India. NABARD also performs a regulatory role, although this bank features among the suppliers of bulk credit too.

The apex bodies, also called the wholesalers, are those who supply bulk credit for onlending purposes. Some of the major institutional sources of bulk credit to the Indian micro-finance sector consist of

- NABARD, Mumbai;
- Small Industrial Development Bank of India (SIDBI), Lucknow;

- Rashtriya Mahila Kosh RMK, a government initiated NGO under the department of Women and Child Development;
- Housing and Urban Development Corporation (HUDCO), New Delhi;
- Housing Development Finance Corporation (HDFC), Mumbai;
- Friends of Women's World Banking (FWWB), Ahmedabad

The next components of the institutional mapping are the intermediary agencies, which comprise of banks of various kinds-scheduled, regional rural and co-operative, micro-finance institutions and facilitating NGOs.

**Regulatory Bodies**

Regulatory bodies like the Reserve Bank of India and NABARD, perform the role of officially regulating micro-finance interventions and providing directions for policy. The RBI sees the provision of micro credit by banks as an important instrument for poverty alleviation, particularly in the rural areas, as it raises the productive capacities of the beneficiaries.

In keeping with this philosophy, in February 2000, banks were advised to make micro credit an integral part of their corporate credit plan. Since then micro credit is reckoned as part of the banks' priority sector lending. Separate data about credit flow to women is to be generated by the banks and quarterly reports submitted to the RBI, which will create a separate database for women. Data are to be collected separately for micro credit, for credit to the small scale, medium scale and large scale industries. Already, most banks have started collecting data, as required, from the branches, on a periodic basis.

The data is consolidated at the head offices. A 14-Point Action Plan, which concerns credit for women, was outlined for implementation by banks. By the end of the first year of the implementation of this Plan, factual data would be available on the number of women financed, the amount of credit flow extended to women as also the percentage to net bank credit. It is recommended that the RBI should prescribe a special format for data collection from the banks so that the data is available separately for amounts advanced to women entrepreneurs for micro credit, for small/medium/large scale industries.

NABARD is the primary agency for coordinating and facilitating the extension of rural credit. Commercial banks supplement the efforts of NABARD and co-operative banks, in meeting the credit requirements of rural India. In the year 2000, the RBI set up a special cell to liaise with NABARD and other institutions for augmenting the flow of credit. A Task Force was instituted by the Bank to evolve an organizational and regulatory framework for mainstreaming micro-finance operations.

Prior to that, the SHG - Bank linkage was introduced by NABARD in February 1992, as a Pilot Project to cover 500 SHGs with policy back-up from the RBI. NABARD brought together the SHGs already promoted by NGOs, and the banking system, to provide various financial services. The launch was tried in Karnataka and Tamil Nadu. The pilot phase was followed by the setting up of the Working Group on NGOs and SHGs by the RBI in 1994. This came up with wideranging recommendations on the internalization of the SHG concept as a potential intervention tool in the area of micro-finance.

Totally, under the linkage program, 81,780 SHGs were provided micro credit by banks during 1999-2000 bringing the cumulative number of SHGs credit linked to

114,775 as on 31March 2000. NABARD has also introduced the concept of bulk lending to NGOs in the form of revolving fund assistance to be provided on a selective basis to NGOs who find it difficult to secure loans from banks for on-lending to SHGs.

The amount of such assistance, cumulative upto September 1999 is Rs.110.7 million. NABARD provides 100 per cent refinance assistance to banks at an interest rate of 6.4 per cent per annum for financing SHGs. Cumulatively till March 31, 2000, the bank loans disbursed to the SHGs aggregated Rs.1929.8 million while NABARD's refinance availed of by the banks aggregated Rs.1501.3 million. NABARD has also been providing grants on a selective basis for the capacity building of the NGOs that are involved in microfinance activities.

The cumulative grant assistance sanctioned till March 31, 2000, for promotion and linkage of SHGs, aggregated to Rs.24.8 million, covering 72 NGOs and 16,911 SHGs. Additionally, NABARD also provides other types of support, such as, for training of bank staff, NGOs, and government agencies involved in the area of micro-finance. As on 31 March 2001, the regional spread of SHGs was as follows the cumulative number of SHGs credit linked in the northern region was 3222, in the north-eastern region was 196, in the eastern region was 9298, in the central region was 15256, in the western region was 7983 and in the southern region was 78720.

Totally in India 114,775 SHGs were credit linked as on 31 March 2001. NABARD's credit schemes are gender neutral. Focussed attention to gender issues in credit and support services, is given through the women's cell set up by NABARD where credit is provided in the form of refinance assistance to the banking system in respect of their advances for agricultural, allied sectoral rural cottage and village industries activities.

NABARD has brought out a number of schemes exclusively for the benefit of women- entrepreneurs. These are

- Assistance to Rural Women in Nonfarm Development (ARWIND)
- Support for setting up women development cells (WDC)
- Assistance for Marketing of Nonfarm Products of Rural Women (MAHIMA)
- Gender Sensitization Programs NABARD has also sanctioned a number of credit-linked promotional programs in the form of Rural Entrepreneurship Development Programs (REDPs), Artisan guilds, and Training-cum Production Centres (TPCs), with the intention of enhancing women's entrepreneurial capabilities and settling them into selfemployment and wage employment.

The research study 'Micro-finance Regulation in India', unravels the features of a self-regulatory mechanism in micro-finance in India. The sequel to the report of the National Task Force on Supportive Policy and Regulatory Framework for Micro-finance, had suggested the need to encourage the development of a self-regulatory mechanism and had also defined some of the broad parameters under which such a regulatory mechanism could be given shape.

The role of the government is seen as one that provides an enabling environment and does not proactively direct and control the activities of the participants. The potential objectives and benefits of self-regulation lie in

- Protecting the depositors,
- Offering saving services,
- Gaining access to refinance facilities

— Lower refinance costs,
— Enhancing the image of the sector,
— Helping MFIs build capacity and improve performance,
— Pre-empting regressive governmentintervention.

The report lists the participants that should come under self-regulation as - the SHGs; the Federations of the SHGs; NGO MFIs; NGO facilitators; Co-operatives; other formal MFIs like banks, NBFCs, and LABs. The Self-Regulatory Authority (SRA) is expected to verify if the MFIs are really serving the poor, set accountancy rules, reporting and disclosure requirements, establish prudential norms and performance standards, and accredit the complying MFIs.

It is also expected to give credit ratings and provide deposit insurance financed from contributions of members. The study recommends that even if regulation norms and standards are established by the MFIs themselves, supervision must be done by an independent institution.

**Wholesalers**

Apex bodies are responsible for the disbursement of bulk credit for on lending purposes. Following are some of the major apex bodies operating in the Indian micro-finance sector. It must be noted here that NABARD is also an apex body along with being a regulatory one. As examples, below are outlined the operations of SIDBI and Rashtriya Mahila Kosh (RMK).

SIDBI Foundation for Micro Credit (SFMC), in November 1998. It is a fully functional separate department of SIDBI since January 1999. The Foundation is supposed to assist NGOs and voluntary organizations on soft terms, which would in turn extend micro credit to poor people. As of March 31 2001, the SFMC has 49 MFI clients. It has an outstanding loan portfolio of Rs. 332 million. 150,000 are ultimate clients, out of which 80 per cent are poor and 70 per cent are women.

SIDBI has designed a number of schemes and programs to support women entrepreneurs with financial assistance as well as providing training and extension services. These schemes are, Mahila Udhyam Nidhi, Mahila Vikas Nidhi, Micro Credit Schemes (MCS). Under the MCS fund, support is provided to well managed NGOs to on-lend to rural poor women to set up their own micro enterprises. Wherever SHGs exist, the NGO lends to them, and they in turn advance the credit to their members.

The cumulative assistance under MCS for the period March 1999, aggregated Rs. 307.1 million, channelised through 142 NGOs/ MFIs across 24 states/union territories benefiting over 112,000 rural poor, of which more than 90% were women. In all, 97 programs have been supported with the assistance of Rs. 6.7 million for providing training to 1550 NGOs in the areas of credit usage, management, monitoring and administration on credit programs.

The Government of India established the Rashtriya Mahila Kosh (National Credit Fund for Women), in 1993, with a corpus fund of Rs. 310 million. This initiative was taken since there was a need for a quasi-formal credit delivery mechanism which was client friendly, had simple and minimal procedures, disbursed credit quickly and repeatedly, had flexible repayment schedules, linked credit to thrift and savings, and had low transaction costs.

The Main Loan Scheme of the Kosh aims to provide credit to poor women both in the urban and rural areas for income generation activities (unless specifically sanctioned for other purposes). Specifically, women below the poverty line are eligible for support. The credit facility is channeled through eligible organizations to needy women without the insistence of collateral. During the year 2000-2001, totally in the country, RMK attended to 46,559 borrowers.

The cumulative number of borrowers since inception upto 31, March 2001, was 406,942. During the same period, 861 new NGOs availed of RMK assistance and 354 repeated previous experience. In all, the amount sanctioned by RMK was Rs.1034.73 million and the amount disbursed was Rs. 765.53 million for the same period.

# 7

# Rural Infrastructure Development

Dynamic gains come from higher growth, potentially raising the entire future stream of consumption. Reductions in transaction costs can increase growth rates as well as providing static efficiency gains. ICTs can also spur innovation, which is a key factor in economic growth. The second type of potential benefit comes from reductions in economic inequality, to the extent that such reductions are an agreed-upon social goal, and therefore a social benefit.

The two types of gains may conflict, if growth requires increased inequality, or they may be mutually reinforcing, where broad sharing of the benefits of growth enhances the rate of growth. The role of ICTs in achieving greater economic gains along both dimensions, without having to commit to a particular position on the relationship between inequality and growth.

However, a focus on using ICTs for rural development is, at least on the surface, supportive of reduced inequality along with increased efficiency and growth, to the extent that rural residents of developing countries are relatively deprived in their access to modern communications. On the other hand, if rural ICT access and use is restricted to the privileged component of those societies and communities, inequality within rural areas would be exacerbated.

Of course, benefits that are measurable as increased market-based economic activity, and hence show up in national accounts statistics, are not the only component of development. Development can also include improvements in the capabilities of the population, such as education, health and nutrition, independently of any direct or indirect economic impact.

The ability to participate in democratic decision-making also falls into this category. Broad-based improvements in capabilities can also have positive impacts on long-run economic well being, but this is not a necessary condition for desiring such improvements. The role of ICTs in effecting improvements along noneconomic dimensions is also considered in this monograph. Turning to specific impacts, note that digital ICTs involve the electronic processing, storage and communication of information, where anything that can be represented in digital form is included in the term 'information'.

Thus news, entertainment, personal communications, educational material, blank and filled-out forms, announcements, schedules, and so on are all information. Software programmes that process data (searching, tabulating, and calculating, for example) are also information in this sense, representing a particular kind of intermediate good. Information goods typically have the characteristic that one person's use does not reduce their availability for another person.

Thus, a message or weather news can be viewed by many people, simultaneously or sequentially. Depending on the content of the news or message, different people may place different valuations on the information. Only friends and relatives may be interested in a personal message, all farmers in a district may be interested in local weather news, and so on. The ability to share information among users can impact the feasibility of providing it on a commercial basis. Digital IT dramatically increases shareability of information, and this affects the economics of private provision of information goods and services free riding may lead to suboptimal private provision. The government may therefore provide information goods because they are shareable and non-excludable (pure public goods). The classic example of a pure public good is national defense, but such goods may also be local in nature, such as public parks or law and order.

Of course many local shareable goods can be provided exclusively, in which case private provision is a feasible alternative, in a club-like arrangement. Here, government provision may be justified more on equity grounds than on the basis of failure of private provision. In some cases, government financing through taxes or statutory user charges can be combined with outsourcing of delivery to private providers to achieve both equity and efficiency goals. In the case of rural ICTs, the benefits may come from a mix of privately and publicly valuable information.

Often, private provision is feasible, but neglects the spillover benefits that it creates, in which case government subsidisation may be socially beneficial. For example, primary education has private economic benefits that people are willing to pay for, but it can also have substantial non-economic benefits to the individual and to others in the society (improved understanding, ability to make sound judgments, political decision-making capacity, and so on).

Positive spillovers and shareability may both provide an economic rationale for some degree of government cost-sharing or subsidisation, e.g., for ICT infrastructure. Of course, many kinds of physical infrastructure, such as roads and ports, have similar public good-type benefits.

Additional roles of government that are important to bring out are in redistribution to achieve equity objectives, and in regulation of private activities through licensing and certification. In both cases, the government also uses economic resources, and ICTs has a potential role in increasing the efficiency of government. For both government and private provision, one of ICTs' main direct benefits is in increasing efficiency by economising on resource use in the operations of firms as well as in market transactions.

Information that would otherwise be conveyed through face-toface contact, post, courier, print delivery, telegraph or telephone may instead be communicated in digital electronic form via the Internet, saving time as well as physical resources. Efficiency gains from Internet use are not automatic the telephone, in particular, is an efficient means of communication for many types of information. ICTs also require new investment, so the benefits of trips, time and paper saved must be weighed against the costs of installing and maintaining the new infrastructure.

Yet another significant cost, often more than the infrastructure, is that of training and transition to new work practices. In this case, however, one can argue that there are lasting spillovers through the creation of human and organisational capital. Efficiency benefits of ICTs are not restricted to the communication itself. Information technology (IT) can improve the efficiency of the telephone network, and they can make it possible to track and analyse commun-ications.

Word processing, maintaining accounts, inven-tory management, and other such activities that may not require long-distance communications are also made more efficient by IT. IT also improves the accuracy, updating, retrievability and backing-up of data required for a range of personal, business and governmental activities. Experience with Internet use in developed countries suggests that information exchange related to the completion of market transactions is especially valuable. The ability of IT-based communications to bring together buyers and sellers more effectively represents major potential gains.

These gains can come about through lower search costs, better matching of buyers and sellers, and even the creation of new markets. The successes of auction and employment websites in the US illustrate these gains. In the rural Indian context, farmers selling their crops and buying inputs, parents seeking matrimonial alliances for their children, and job seekers are all potential users of Internet-based matching services.

Depending on the nature of the market, including type of good or service, mobility of participants and the length and value of the transactional relationship, the overcoming of geographical barriers can be significant. Thus, in the Indian context, job markets have seen the greatest benefits in shrinking distance, followed by marriage and then crop markets. Efficiency gains of ICTs can also come about through the enabling of new goods and services. In many cases, the new good is related to something available earlier, but is presented in a form that reduces costs and expands the size of the market.

Recorded music is a mass-consumption item, whereas only a small minority of the population could afford or have access to live performances by the highest quality musicians. Educational material is another example where recording and duplication can replace more expensive, skilled-labour-intensive alternatives for delivery. The possibilities for interactivity with IT-based educational materials illustrate the advantages of digital ICTs over older technologies based only on recording and duplication and analog broadcast communications. Interactivity also implies personalisation, in that an individual can select the precise content that he or she wishes to see. This feature also distinguishes digital IT-based content from what was available through previous technologies. Finally, the sheer volume of information that is accessible through IT is much greater than before this also allows new kinds of services to be provided at a cost that is affordable to larger segments of the population.

The potential static efficiency benefits of ICTs, but the direct dynamic benefits in terms of higher growth are harder to identify empirically. Of course, if IT economises on current resources, more is available for investment, which can increase growth. If ICTs increase the efficiency of education delivery to the broader population, this investment in people is also likely to lead to higher growth.

IT may also have positive impacts through impacts on the innovation process. For example, IT can make innovation easier by allowing simulation and lowcost testing of new designs or searching through possible chemical compounds for beneficial drugs. Furthermore, IT may speed the diffusion of innovations through better communications, which may stimulate further innovation. One possibility is that ICTs may increase growth by reducing transaction costs. This is analysed by Singh as follows. In his model, a reduction in transactions costs increases the number of intermediate goods that are produced and in turn this number influences growth.

This is because the productivity of intermediate goods is increasing in the number of varieties, representing a type of complementarity in production. While there is no long-run growth in the model, it can be added in through learning or labour force growth. The steady state of the economy, however, depends crucially on the level of transaction costs. Transactions costs may arrest the process of development in the sense of keeping the number of varieties of intermediate goods at their initial level and/or reduce their long run level, thus reducing short-run growth and long-run productivity and welfare. An important barrier to realising the economic benefits of ICTs is the oftensubstantial up-front cost of investment in new infrastructure—both hardware and software. In developed countries such as the US, large potential customer bases and efficient capital markets help overcome this barrier.

Hardware and software designed for developed country markets can easily be adapted to serve higher income consumers in developing countries, but this leaves out the majority of the population in developing countries. In this scenario, one potential consequence of IT use is an exacerbation of inequality, as only higher income groups enjoy its benefits—this is the so-called 'digital divide.'

On the other hand, because government-provided goods and services, including redistributive transfer payments, are often aimed at lower income groups, to the extent that ICT use can increase the efficiency and effectiveness of government, the benefits of IT will be more widely spread, partly reducing 'digital divide' concerns. However, achieving these benefits requires more than just internal use of IT beneficiaries of government services must be able to access IT resources also. While governments may invest in such front-end interfaces with citizens, the cost of doing so for governments in developing countries may be prohibitive. Such governments typically already have difficulties in raising sufficient resources through taxes and user charges. Again, the equipment costs may be compounded, or even dominated, by the training and other process transition costs.

While successful examples exist of conventional implementation of 'egovernance' initiatives, exclusively through government action, there is a conceptual alternative. This comes from recognising the fact that citizens typically incur private costs (often substantial) in availing of government-provided services.

If the use of IT can reduce such costs, even low-income individuals may be willing to pay at least some fraction of the cost savings, and there is scope for private provision of intermediate services that reduce the cost of access to government. Of course, this idea is not specific to IT private intermediaries already help in filling out forms, getting access, and so on. One difference that IT can make is in reducing costs even further, often by an order of magnitude. In broad terms, IT changes the scope and nature of intermediation.

Private providers may therefore have a role in delivering IT-based information services that are complementary to government services, as well as in providing conventional private goods and services. However, the private individual benefits that determine the prices charged by private providers may not reflect the overall social benefits of provision. As discussed earlier, these may include benefits such as greater awareness and participation in the political process.

In such cases, there may be a role for government subsidisation of private provision. This assumes that government provision is likely to be less efficient than private provision, which seems to be true in some cases in developing as well as developed countries. In either case, richer information flows can increase the transparency with which the government operates, thereby promoting better monitoring, and potentially—depending on whether electoral and legal institutions are effective—greater accountability. The ultimate payoff is more efficient delivery of government services.

Looking at the case of India, in cities and larger towns, cyber kiosks have already begun to proliferate. Urban population densities, income levels, cultural attitudes and telecom infrastructure all seem to be sufficient for the commercial success of these enterprises. The falling cost of hardware and the availability of a variety of English language software have also supported this trend.

Finally, the government's belated opening up of Internet service provision to competitive entrants has been a crucial supporting development. In non-Internet IT-related services, IT education has clearly taken off in cities as well, inspired by India's success in software exports. In rural areas and smaller towns, however, the various demographic and socioeconomic factors such as income levels, cultural attitudes, and geographic and social fragmentation may not be present in configurations that would easily enable the diffusion of commercial access to various IT-enabled services.

Furthermore, the market power of traditional rural intermediaries may act as a barrier to partial innovations in how matching of buyers and sellers is conducted. Finally, vernacular language requirements and different demand patterns imply the need for software that is tailored for fragmented rural markets. All these issues are discussed in more detail in this monograph.

## Rural ICTs

One can examine the potential for rural IT use from both the supply side and the demand side. On the demand side, an overview of the potential benefits that IT can bring to these populations, if the implementation is successful. We begin with the demand side, as a way of motivating the supply side issues. Potential sources of demand for IT-based services can be framed in terms of a simple flow diagram representing the decisions of rural households. A typical household as engaged in farming or related agricultural occupations though this is not true for all of them, farming probably constitutes a significant activity for the majority of rural households.

Input decisions include material inputs such as seeds, fertiliser and pesticides; and capital inputs such as tractors and land (whether through purchases or rentals); as well as the credit required for such purposes. Much of the focus analysis for potential applications of ICTs will be on market transactions for inputs. In all cases, there is a potential for benefiting through improved information about prices, quality and availability.

Labour is also an important input, but the labour market has special characteristics that reduce the importance of such information in rural contexts. This contrasts with urban settings, where online job posting sites have been quite successful. Farming operations include decisions with respect to quantity and timing of inputs. A crucial aspect of agricultural operations is risk management, as both the weather and pest incidence are extremely variable.

Ex ante decisions in the face of uncertainty, as well as ex post responses to realisations of uncertainty are both important. Predictive and technical information are both important for agricultural operations.Increasingly, producing for sale also requires knowledge of quality requirements in different markets. This was emphasized by ITC managers as an important motivation for, and benefit from its e-choupal initiative. Selling of produce provides income for consumption, investment and inputs, as well as generating information about demand that can influence future input and production decisions.

Non-production decisions of households can be roughly divided into pure consumption activities, as well as human capital investment in education and health. The boundary is fuzzy, since even basic food consumption has nutritional impacts and therefore a human capital component.

These activities generate impacts on operations, since they affect the quality of decision-making, as well as household labour inputs indicate this with the vertical arrows. Health and education are obviously two of the most important areas where rural India suffers severe deficits compared to generally acceptable societal standards. While poorer households may be severely limited in their consumption, a significant proportion of rural households in India have incomes sufficient to support some discretionary consumption.

Finally, the rightmost box captures household saving, again something that is not feasible for all rural households, but nevertheless an activity of growing importance -specifically in the case of financial saving. Saving may be for consumption smoothing, investment, or precautionary reasons. In India, bank nationalisation and postal savings schemes have provided rural households with convenient opportunities for financial saving.

Typically, nationalised banks have rural branches, but they do not have levels of efficiency (including low enough transaction costs) sufficient to effectively provide credit to small farmers or non-farm rural households. Microcredit institutions are a well-known alternative, but have limits on scalability. ICTs have a potential role to play in enhancing the provision of rural credit this possibility.

Even this simplified and summary picture of rural households' economic activity illustrates that they engage in a broad range of transactions and decisions with economic impacts. What is noteworthy besides this complexity is that many decisions are made with very limited information, and that market interactions are often subject to high transaction costs, due to imperfections and asymmetries in information, as well as high transportation costs, inefficient intermediation and time delays. High transaction costs will always prevent marginal transactions from being undertaken; in extreme cases, the market may fail to function at all. Given this scenario, the role of IT can be understood in terms of reducing transaction costs, as well as improving the efficiency of decision making within households.

## Organisational and Policy Issues

Turning to the supply side, the various stages of decision-making and delivery of ICT-based services in terms of a typical value chain. At each stage of the chain, the ICT components include a mix of hardware, software and services. For example, an Internet kiosk would have a computer, printer, web cam, modem, power back up, and software to enable standard Internet browsing, as well as handle specialised tasks such as education in the local language, agricultural information, e-governance and entertainment.

The organisational structure necessary for the delivery of rural IT services typically requires commercial goals of profitability to be built in at some level, if scalability and sustainability are to be achieved. In the absence of commercial goals, the organisation's incentive structure is unlikely to be financially sustainable without continual external infusions of support.

The organisational models of experiments such as the MSSRF in Pondicherry and DHAN in Tamil Nadu both rely on ongoing external funding. The reason for this is an emphasis on "developmental" goals over financial sustainability. As will be clear, however, all rural ICT projects require some degree of initial subsidisation, and some ongoing cross-subsidisation the difference in the organisational models therefore lies in the balance between pecuniary and non-pecuniary motives as sources of incentives for efficient action. There is no unique answer, but it can be argued that a proper role for commercial incentives is essential if positive impacts are to be long-term and large-scale.

At the other extreme, a standard corporate structure is used ITC is the most significant example of this approach. Aksh, an optical fiber company, and EID Parry, a processed food company, have both been involved in rural ICT efforts, though not as comprehensively as ITC. It is clear that, for scalability, some minimum size of the organisation is required, to spread the fixed costs of administration efficiently.

In addition, there are fixed costs of innovation that can be spread more effectively across a larger organisation. Note that the organisation referred to in this case is not necessarily the whole company, but could be the division that implements the rural ICT service project. A variant on the corporate model is where there is a producer organisation that controls the project, as in the case of the Warana sugar cooperative in Maharashtra.

By far the most common organisational model that is emerging is a hybrid one, combining non-profit and profit motives. For organisations that are dedicated specifically to rural ICT-based services delivery, social goals can be incorporated by vesting controlling ownership of the corporation in a non-profit entity that has an explicit social focus. This model has been followed by n-Logue, TARAhaat and Drishtee, for example.

A variant on this involves controlling ownership through the public sector, typically the state government, though in smaller-scale cases, a rural local government may be the main stakeholder.

Even in the case of existing for-profit corporations with broader businesses, social goals may enhance reputation, meet corporate social responsibility guidelines, or otherwise be consistent with the mission and values of the organisation. In other words, including social goals may make good bottom-line business sense. In the case of ITC, this is clearly a factor in their efforts, though the supply chain efficiencies that are realised are critical to the expansion and continuation of their efforts.

In another case, Aksh was not immediately able to realise commercial objectives in its support of kiosks in Rajasthan, and the reputational enhancement alone was insufficient to provide quick sustainability of that ICT initiative. Nevertheless, Aksh has used a recent increase in state government interest and consequent possibilities for e-governance to expand its efforts in Rajasthan, as well as enter Andhra Pradesh the commercial motivation comes from the provision of fiber optic cable and cable TV delivered over that cable, with rural Internet services being an "add-on".

For all types of organisations, building the right capabilities requires some effort. Creating what amounts to a brand new infrastructure for rural ICT service delivery requires a broad mix of skills, and finding talented and trained people who can be effective in a role that mixes entrepreneurial tasks with corporate line responsibilities, all in an unfamiliar rural environment, can be a challenge. In the case of new efforts such as n-Logue, TARAhaat and Drishtee, this organisation-building task as a major challenge.

One significant aspect of this challenge is finding skilled people willing to work in rural areas. In one case, the field managers of an organisation were more comfortable in an urban corporate setting, rather than dealing with villagers and "small-town" people. N-Logue found that hiring in cities was very expensive, and has tried to identify and train middle-level field managers from the villages and towns in which it operates kiosks.

For existing organisations, the problem of hiring, training and organisationbuilding is somewhat mitigated—nevertheless, there is apparently a need for high level direction to ensure that resources are allocated for the effort, including human resources. In the case of ITC, the heavy use of management trainees to interact with and support the e-choupal effort appears to be a crucial success factor, providing a continuous supply of fresh young talent, and ensuring that over time, the larger organisation will have a deep understanding of the role of e-choupals in ITC's global business.

In the case of smaller, non-commercially oriented efforts, of course, these organisational challenges are less significant. Running a few kiosks may involve additional paid staff, but a major source of the required skills comes from a rotating pool of volunteers. One important question that will only be answered with time is the long-run sustainability and scalability of this volunteer model. The level of external resources required to maintain it may also be a concern.

Where the government is involved as the controlling organisation, the experience in India suggests that a centralised and personalised model is adopted. Typically, someone high up enough in the government hierarchy develops an enthusiasm for using ICT for e-governance or other services, and this enthusiasm translates into a "big push."

Also typically, these efforts do not involve essential modifications in governmental structures or processes, and hence there is not a systemic innovation that can lead to sustainability. This is a common feature of several rural e-governance efforts in particular, as described in Dossani et al and Parthasarathy et al.

One response to the significant challenges of collecting the necessary talent and skills is to enter into partnerships with other organisations that may provide specific pieces of the overall package that is needed application software, content, maintenance services, technology, marketing, and so on.

In particular, the ICT components in the middle three stages of the value chain may require partnerships, though less so for commoditised components such as PCs, and more for infrastructure pertaining to connectivity and applications. These partnerships can pose their own problems, particularly in goal alignment and consequent performance monitoring requirements.

The relatively unsuccessful partnership between Drishtee and Aksh to manage kiosks in Rajasthan is an example of these problems. In cases such as this, the different resource scales of the partners also can make collaboration difficult to sustain. In several cases, various informal as well as formal partnerships have worked better, and have involved various mixes of NGOs, government and private sector entities.

In only one case, namely n-Logue, has there been a significant, ongoing university collaboration, in that case, IIT Madras. This collaboration has spanned infrastructure as well as applications. Given the nature of these initiatives, with substantial room for organisational and technical innovations, it is noteworthy that other technical institutes and management schools have not been more involved.

Of course all the organisational issues involved in setting up rural ICT efforts are well recognised in management writing and experience what is important is to recognise their criticality in a setting that has more typically been the arena of pure 'social service' entities such as governments and NGOs. It is also critical to recognise that the organisational innovation required in this case is an order of magnitude greater than the task of selling consumer goods in rural areas, something that is now accepted as a given in India. The greater complexity and variety of the services being delivered through ICT is the root cause of this difference.

Government policy has been reasonably accommodating from the perspective of organisational issues. NGOs and hybrid organisational forms are well-established methods of bringing together the components of the rural ICT value chain, and individual government officials and departments have been supportive and cooperative at various times and places.

The final stage of the value chain in Figure 2. refers to human resource management (HRM) and customer relationship management (CRM). In this context, these more general terms take on specific focuses. Training of rural kiosk operators, whether they are formal franchisees or independent farmer operators, becomes a key aspect of the delivery model. Training the field personnel at various levels (village and district hub) is also critical.

Gathering customer information on usage patterns (nature and timing of use), revenue streams, responsiveness to pricing, social acceptance, and so on is also vital, as these are brand new markets in terms of the nature of service delivery. Furthermore, being able to respond to this information with appropriate and timely changes in strategy is also necessary for successful implementation. In the case of n-Logue, "local service providers," as partners-cum-franchisees, handle some of these ongoing functions, with initial support provided by n-Logue. In other cases, this ongoing management is handled without partnerships.

## Infrastructure Development

The second stage of the supply chain in Figure 2. concerns access to electric power and Internet connectivity. In both cases, a major constraint on ICT initiatives is the failure of the public sector to deliver adequate power and telecommunications to rural India. This is no longer because these technologies are too expensive or sophisticated for a poor country—the failure is purely institutional. Privatisation has helped in the case of telecommunications, as has technological change. In particular, the widespread adoption of cellular phones has been driven by the opening up of this sector to competition almost from its inception.

Of course, innovation in digital communications technologies is the foundation of all rural ICT-based service delivery. While conventional telephone connectivity has often proved inadequate for Internet access in rural areas, because the quality of existing voice lines is too poor to sustain data transmission, several innovations provide alternatives that are likely to be cost effective. These include wireless in local loop (WLL), fiber optic cables, and high-powered versions of Wi-Fi and Wi-Max.

The Internet boom in the United States clearly played a role in pushing down costs and speeding innovation in fiber optics and wireless transmission. In some cases (ITC in particular), VSAT (very small aperture terminal) satellite connectivity has been used for Internet access, but it is not very cost effective. The major challenges for connectivity are likely to be regulatory, having to do with interconnection to the main network, and with maintenance, rather than with the fundamental technological choices and implementation.

Combinations of a fiber optic backbone (which is now extensive in India), and wireless technologies for last-mile access (specifically, newer versions of Wi-Fi and WLL ) are likely to emerge as cost-effective solutions, provided the government is able to implement socially desirable policies for investment as well as interconnection charges. The main thrust of the report of

Dossani et al is to argue for using Universal Service Obligation (USO) funds collected from telecommunications providers to finance improved rural ICT access. This perspective is a valuable one.

Digital convergence in ICTs means that separating voice from data in formulating policies for rural access no longer makes sense. Similarly, calls to focus on mobile phone access rather than Internet kiosks neglect the convergence in technologies as well as the different functions served by the two access modes.

The government and its regulatory agency, TRAI (Telecoms Regulatory Agency of India) have announced a move toward unified regulation, but the roadmap is unclear and progress slow. One problem is that the interests of the large government-owned incumbent BSNL are protected through restrictions, e.g., on the use of VoIP. Electric power is in many ways more of a problem than telecoms connectivity, and this is true throughout India. The reason, as noted, is a serious institutional failure.

The production, transmission and generation of electricity in India are seriously inefficient, as a consequence of poor organisational incentives and vested interests in the traditional public sector. For rural ICT access, the lack of power for long periods seriously hinders accessibility. Battery backups are an essential part of rural Internet kiosks, though they are a very partial solution to the lack of reliable power supplies since they cannot provide sustained power during long power outages. The need for heavy duty battery backups also adds significant capital and maintenance costs for these kiosks.

Of course, this is a microcosm of the problem faced by all electricity consumers in India -they are forced to use inefficient, small-scale backup power supplies, because of the total inefficiency and unreliability of what should be much cheaper large scale provision of electricity. Solar technologies may be more promising in the near future they are already in use in existing rural IT efforts, particularly that of ITC.

Realistically, electricity supply in India will not improve dramatically any time soon, and innovation in solar panels, battery technology and electricity storage may eventually provide attractive alternatives for rural India in general. The third stage of the supply chain is the most straightforward, because of the standardisation of components of desktop computing and peripherals, rapid technological improvements, falling costs of production, and, most recently, price reductions resulting from changes in tariffs on imported hardware. It is now possible to fully equip a singlecomputer rural Internet kiosk for less than Rs. 50,000, including CD drive, printer, scanner, power backup, and web cam.

Potentially, the highest cost component is the operating system, since Windows has enjoyed a virtual monopoly on the desktop. However, Microsoft does have some concessional pricing for socially oriented developing country initiatives, and this helps to reduce costs. The continued development of Linux as an open-source alternative for the operating system also helps to bring costs down, both directly, and through providing competition for Microsoft.

In fact, government adoption of Linux and the development of browser-based applications can both hasten the use of Linux. Large scale adoption of Linux by companies such as n-Logue might also provide an impetus for domestic entrepreneurs to develop Linux-based applications targeted for regional rural markets. Browser-based applications may also make "network appliances" more feasible.

The operating system is still typically in English, but as long as simple visual (using icons) and keyboard drills can take kiosk operators to local language applications and content, this is not a substantial usage barrier. Kendall and Singh find, in a sample of n-Logue kiosks, that an operator's having a college degree or even having completed high school does not significantly increase kiosk revenue the minimum level of education needed to be successful is not very high, provided adequate training is provided.

One can conclude that this stage of the supply chain is easiest to implement, with a highly standardised, almost cookie-cutter approach—although ongoing maintenance can be a challenge. The major business decision is whether to have more than one computer per kiosk, but experience suggests that one is sufficient for almost all situations, at least in the beginning. The next stage of the supply chain, namely applications, presents more challenges. The range of possible applications is vast. Many IT-based services require non-IT logistics or processes as complements.

Availability of local language software becomes more of a constraint. There is much more variation across localities, not just regions. Delivery of services or development of content often stretch the resources and expertise of the primary provider, and require varied partnerships or other contractual relationships. Local language content development can be hindered by a lack of software and hardware standards.

Deciding the sequencing, scope and sophistication of various applications can be a major challenge, since many of the services are being offered for the first time, or are being delivered in novel ways that challenge existing institutional frameworks and relationships. However, one of the benefits of the numerous rural ICT experiments that have been conducted throughout India is that the kinds of applications that are valuable to rural households have been identified and refined.

As one would expect, computer games are popular with children, and some kinds of communication and information retrieval are highly valued. Word processing is often needed by children and adults, and digital photographs are also very much in demand. One can generalise somewhat, to say that basic digital applications, often taken for granted in developed countries, are the basis for any rural IT kiosk's financial sustainability.

## Rural-urban Interactions and Livelihood Strategies

The division between 'urban' and 'rural' policies is based on the assumption that the physical distinction between the two areas is self-explanatory and uncontroversial. However, there are three major problems with this view. The first is that demographic and economic criteria used to define what is 'urban' and what is 'rural' can vary widely between nations, making generalisations problematic.

A second problem is that of the definition of urban boundaries. In Southeast Asia's Extended Metropolitan Regions, agriculture, cottage industry, industrial estates, suburban developments and other types of land use coexist in areas with a radius as large as 100 km, where the high mobility of the population includes circular migration and commuting.

In sub-Saharan Africa, agriculture still prevails in peri-urban areas. However, elsewhere significant shifts in land ownership and employment patterns are taking place, often at the expense of both rural and urban poor people. The third problem in the definition of the boundaries between 'rural' and 'urban' areas is the fact that urban residents and enterprises depend on an area significantly larger than the built-up area for basic resources and ecological functions. In general, the larger and wealthier the city, the more its industrial base and its wealthy consumers will draw on such resources and ecological functions from beyond its surrounding region.

The concept of a city's ecological footprint was developed to quantify the land area on which any city's inhabitants depend for food, water and other renewable resources such as fuelwood, and the absorption of carbon to compensate for the carbon dioxide emitted from fossil fuel use. The concept makes clear the dependence of any city on the resources and ecological functions of an area considerably larger than itself.

Internal migration is often seen as essentially rural-to-urban and contributing to uncontrolled growth and related urban management problems in many large cities in the South. This has resulted in many policies to control or discourage migration. While migration restriction is infrequent, many countries have sought to make cities relatively inhospitable, for example bulldozing informal low-income settlements, or making it difficult for new migrants to secure property rights to land or access to public services.

These measures generally have little impact aside from lowering welfare, especially for the poor. In fact, most of the growth in urban population is due to natural population increase. Since rural to urban migration is fastest where economic growth is highest—as migrants tend to move to places where they are likely to find employment opportunities—it is not in reality as problematic as it is made out to be.

Despite widely-held beliefs that flows are always rural-to-urban, migration from the urban to the rural areas is increasing. This type of movement is often associated with economic decline and increasing poverty. In sub-Saharan Africa, significant numbers of retrenched urban workers are thought to return to rural 'home' areas, where the cost of living is lower.

Seasonal waged agricultural work in rural areas can also provide temporary employment for low-income urban groups. Temporary and seasonal movement such as this is not reflected in census figures, and can make 'static' enumerations of rural and urban population unreliable.Complexity in migration direction and duration is matched by that in the composition of the flows, which reflect wider socioeconomic dynamics.

Although regional variations can be important, the number of migrant women has increased in many countries in the South. The age and gender of who moves and who stays can have a significant impact on source areas in terms of labour availability, remittances, household organisation and agricultural production systems.

Household membership is usually defined as 'sharing the same pot', under the same roof. However, the strong commitments and obligations between rural-based and urban-based individuals and units show that in many instances these are 'multi-spatial households', in which reciprocal support is given across space. For example, remittances from urban-based members can be an important income source for the rural-based members, who in turn may look after their migrant relatives' children and property.

These linkages can be crucial in the livelihood strategies of the poor, but are not usually taken into consideration in policymaking.

Exchanges of goods between urban and rural areas are an essential element of rural-urban linkages. The 'virtuous circle' model of rural-urban development emphasises efficient economic linkages and physical infrastructure connecting farmers and other rural producers with both domestic and external markets. This involves three phases:

1. rural households earn higher incomes from production of agricultural goods for non-local markets, and increase their demand for consumer goods
2. this leads to the creation of non-farm jobs and employment diversification, especially in small towns close to agricultural production areas
3. which in turn absorbs surplus rural labour, raises demand for agricultural produce and again boosts agricultural productivity and rural incomes.

In Durban (South Africa), maintaining both an urban and a rural base also provides a safety net for low-income city dwellers in times of economic hardship or political violence. However, housing and rural development programmes do not acknowledge such multispatial, extended households: eligibility to subsidies and grants is based on the size of the co-resident household, and the funds can only be used in one of the two locations. Since urban housing subsidies are more widely available, this may encourage urban-based members of multi-spatial households to cut their rural links.

However, spatial proximity to markets does not necessarily improve farmers' access to the inputs and services required to increase agricultural productivity. Access to land, capital and labour may be far more important in determining the extent to which farmers are able to benefit from urban markets.

In Paraguay, despite their proximity to the capital city, smallholders' production is hardly stimulated by urban markets as their low incomes do not allow investment in cash crops or in production intensification to compensate for the lack of land. Patterns of attendance at periodic markets also show that distance is a much less important issue than rural consumers' income and purchasing power in determining demand for manufactured goods, inputs and services.

## SECTORIAL COMMUNICATION

The growth of urban agriculture since the 1970s has long been understood as a response to escalating poverty and rising food prices or shortages, often exacerbated by structural adjustment and economic reform. Recent research shows that its nature may be changing, and that at least in low income nations, a significant proportion of high and middle-income urban farmers engage in commercial production.

More needs to be known on how this may affect access to urban food markets for producers, especially smallholders, from surrounding rural areas. The increase in non-agricultural rural employment, or deagrarianisation, is an ongoing process in most countries in the South. There are several reasons for this. Amongst them, environmental degradation, population growth and land subdivision make it difficult for large numbers of farmers in many regions to rely solely on agriculture.

Access to nonagricultural rural employment is mediated by culturally-specific formal and informal networks which may be based on income, political and/or religious affiliation, ethnicity, household type, gender and generation. This can constrain some groups' access to the opportunities provided by deagrarianisation and occupational diversification.

Since the 1970s, comprehensive rural-urban development frameworks have been formulated as an explicit attempt to promote rural development, and with the implicit aim of curbing migration to large cities. Integrated Rural Development has contributed the view of rural development as holistic and multifaceted, and including non-agricultural as well as agricultural activities. However, it has rarely included explicit urban components, and whenever a spatial dimension is included it is usually limited to marketing functions.

Other attempts take urban centres as their starting point. In the 'urban functions in rural development' (UFRD) approach, the strategy for promoting rural development is to develop a network of small, medium-sized and larger centres each providing centrally located functions (such as services, facilities and infrastructure) hierarchically organised. Rural development is expected to be stimulated by filling in the supposedly missing functions (for example banking services) through selective investment in rural towns. Translating this model into practice has been problematic for three main reasons:

1. 'urban functions' are assumed to benefit the entire surrounding region and all rural households irrespective of social and economic status: issues of access and control are not considered
2. the methods for selecting key towns for investment were not clear, and tended to focus only on the attributes of the towns themselves with no consideration of the rural potential
3. the model is based on generalisations which do not account for the rich variety in the roles of urban centres, which are determined by both the rural and regional context.

The underlying conceptual problem is the assumption that it is an absence of 'central places' that constrain development, rather than factors such as ecological capacity, landowning structures, crop types and control on crop prices or access to markets, all of which in turn are shaped by rural-urban interactions within the specific regional context.

A third position on the role of small towns in rural development can be defined as 'intermediate'. Drawing on empirical case studies from Africa, Asia and Latin America, it shows that universal generalisations and prescriptions, which form the basis of most spatial planning models, are not valid. Centralised policies which do not take into account the peculiarities and specifics of small towns and their regions, may not be efficient.

Real decentralisation of decision-making with investment and resource-raising at the local level may allow the articulation of local needs and priorities and stimulate both rural and urban development. However, wider socioeconomic issues such as inequitable landowning structures and government crop purchasing policies and taxation are also likely to affect small towns and, by extension, migration to larger cities.

The positive impact of rural-urban linkages on rural livelihoods is summarised in the 'virtuous circle', where rural and urban development are mutually dependent and integrated. However, rural-urban linkages should not be assumed to be beneficial in all circumstances. In some cases, they can increase inequality and the vulnerability of those groups with least assets.

Especially where land ownership is highly unequal, government policies and subsidised credit institutions set up in small towns tend to benefit already privileged urban elites and large farmers. When inputs and services for agricultural development are locally available, those small farmers who cannot afford to buy them tend to lose their land to large farmers who in this way reinvest the profits from their increased production.

Non-agricultural rural employment can be an 'accumulation strategy' for farmers with assets and access to urban networks. For these groups, profits from urban-based activities are often reinvested in agricultural production, resulting in capital and assets accumulation. For other groups, however, engaging in non-agricultural rural employment may be determined by lack or loss of land, capital or labour.

Moreover, social marginalisation can limit access to non-agricultural activities, and individuals and households with little access to social networks, such as in many instances woman-headed households or widows living alone, may be forced to find employment in unprofitable occupations as a 'survival strategy'. The least remunerative of these activities do not reduce vulnerability and may rely on excessive extraction from the natural resource base.

Migration as a livelihood strategy is also mediated by access to assets. Those who move tend to be young, physically fit and often better educated than average, and have access to urban-based social networks. The elderly and the poorest people do not usually migrate, and labour availability in peak agricultural seasons can become scarce. Over time, migration may erode village-based networks as migrants become part of urban-based networks, and remittances tend to decrease.

Migrant women tend to send higher remittances to home areas, although an important reason for moving is often to escape family constraints. The renegotiation of gender roles resulting from women's migration is however not always reflected in an increase in control over their own remittances.

In many cases, women who stay in home areas also have limited control over remittances sent by male relatives. This partly reflects culturally specific gender relations, but also in many cases women's lack of access to assets. Depending on specific locations and groups, projects facilitating the productive use of remittances may have the potential to contribute to poverty reduction.

Rural-urban linkages can follow quite different paths. For local and project research, it is important to identify bottlenecks which prevent positive interactions. This can be done through the analysis of the patterns of different flows and subsequently through the analysis of their combined impact on rural-urban linkages. This holistic approach requires that both the rural and the urban components of the flows' trajectories are included in the analysis. For example, migration should be examined in terms of both labour supply and demand. In flows of goods, access to market price information and to the actual marketplaces should be considered.

The trajectories of the flows are not usually limited within the regional boundaries of a town and its hinterland. Migrants can go to a variety of places, including international destinations, and goods and services can be sold and purchased in many different locations. From the perspective of a rural household, the pattern of flows is thus more likely to resemble a network involving multiple linkages with a number of villages and towns, rather than revolving around a single urban centre.

The focus should therefore be on regional networks rather than on relations between a single village and a single centre. The initial demarcation of these networks can be based on the existing flows of goods and people between settlements.

Secondary data are not a reliable source of information, especially for population flows, since circular and temporary migration are not usually recorded in censuses nor in annual household registration data. The combination of participatory methods, small-scale household surveys and interviews with key informants is likely to be the most efficient approach for local, project related research.

Participatory methods can help in the initial mapping of the flows, and in focus group discussions on the reasons for the patterns identified. Stratified household surveys allow a clearer understanding of how the nature of local rural-urban linkages affects the livelihoods of different groups. Key informants can provide useful information on policies and practices which may not directly emerge from interviews with village respondents.

Why should natural resource policies and projects take into account rural-urban linkages? A first reason is that understanding the scale and nature of rural-urban linkages is essential in order to locate rural livelihoods and the rural economy within the wider regional context. Positive rural-urban interactions and the 'virtuous circle' of development are fostered by backward and forward linkages between agricultural production and industry and services.

However, policies encouraging these mutually reinforcing linkages need to overcome the traditional separation between rural and urban planners. They also need to avoid generalisations and be grounded in the specifics of the regional context. Political and administrative decentralisation involving real decision-making power and financial autonomy is more likely to overcome planning dichotomies and to identify local needs and priorities.

This needs to be supported by the national government, which should also provide infrastructure and basic services. Incentives for localised and diversified foreign investment would help to avoid the weaknesses of investment in single resources and crops. Balanced rural-urban regional development requires an equal distribution of benefits among the rural population, since increases in rural household income and expenditure are the springboard for the expansion of many urban-based enterprises. Inequalities in access to assets are the main reason why there are very few actual examples of 'virtuous circle' development.

To strengthen the poverty reduction element of natural resource policies and projects, the potential of rural-urban linkages in widening choice and options should be taken into account. Increasing people's options also involves recognising the diversity of needs and priorities within any low-income population, not only because of different income but also because of gender, age and ethnicity.

# 8

# Developmental Structure of Rural Marketing

Understanding the characteristics that make the people and the market in rural India unique can help corporations to enter this market with success. The key characteristics define the term rural, determine the amount and flow of income, and determine the types of products and packages that are typically used in rural India. Seventy percent of India's population, or approximately 700 million people, live in rural areas. As of the 2000 census, this equates to just under 2.5 times the population of the U.S. A location is defined as rural if at least 75 percent of the population is agrarian. With such a large number of potential consumers, it is clear why multinational corporations would like to successfully penetrate the rural Indian market.

With an average income equivalent to $42 per month, rural Indians have a very low disposable income. Most rural homes have minimal storage space and no refrigeration. Very few people own or have access to cars. As a result, rural Indian purchasing habits tend to be of an "earn today, spend today" mentality. Rather than buying in bulk, which would mean paying more for a large quantity upfront, rural Indians tend to buy what they need for short segments of time. These factors result in consumers buying products locally, as well as on a daily basis.

In addition to the fact that income levels are low, rural incomes also vary greatly depending on the monsoons. When a monsoon hits, this devastates the livelihood of most rural consumers because they are dependent on agricultural work for income. Corporations are also directly affected because this makes it difficult to predict demand. Before a company considers entering the rural market, understanding the types of products and packages that rural Indians typically use is crucial.

For example, urban Indian consumers would typically use toothpaste for brushing their teeth, while most rural Indians prefer using tooth powder. As a company seeking to enter India's market with an oral care product, this would be an important fact to know and consider during both the product and package development stages.

Similarly, Hindustan Lever Ltd. (HLL), the Indian subsidiary of Dutch-based Unilever, discovered that rural Indians tend to use the same soap for washing everything from hair to their bodies to clothing. Because HLL manufactures products including various soaps and detergents, HHL product and packaging development processes have taken this rural habit into account by designing all-in-one soaps. By taking into account the low disposable incomes and the unique product and package needs of this market, consumer products that are designed and packaged for this market have great potential.

Another key aspect to consider is how and where to produce and package the product. There are several options when a company is attempting to strategically aligning with industry. Companies can partner with an existing Indian company, buy out a local Indian manufacturer of a similar product, or strictly import products into India while keeping manufacturing facilities elsewhere.

## Partnership

The first and best option for aligning with the Indian industry is for the multinational to partner with an Indian company that is already successfully producing and selling a similar type of product. In doing so, the new company can take advantage of the manufacturing facilities and distribution networks that are already in place rather than having to start from scratch.

As a result of India's colonial experience when it was controlled by Britain, many Indians have "...a profound mistrust of foreign brands". By creating a partnership with an Indian company plays down the foreign factor and helps to dispel some of this mistrust. Hindustan Lever is a multinational corporation that has found success with this method of aligning with industry. By partnering with local entrepreneurs who own and manage their own plants, Hindustan Lever is able to manufacture their products with minimal amounts of fixed capital. In these partnerships, the entrepreneurs agree to devote their plant's capacity to manufacturing only Hindustan Lever products.

## Buy-Out

A second alternative for aligning a new industry to enter India's rural market is to buy out a local Indian manufacturer. As with partnering, buying out a local manufacturer gives a company the ability to capitalise on existing manufacturing facilities and distribution networks. The disadvantage is that Indian consumers may view this negatively. Coca-Cola is an example of a multinational corporation that tried buying out a local distributor.

In 1992, Coca-Cola made its second appearance to the Indian market. In an attempt to eliminate its biggest competitor, Coca-Cola acquired Thumbs Up, the local market leader in cola. When Coca-Cola tried to exchange its own brand on the regular Thumbs Up distribution network, Indian consumers looked unfavourably upon Coca-Cola. The company has been struggling ever since.

Additionally, companies can enter India's rural market by importing products from manufacturing locations overseas. Importing has only been a viable means of getting consumer goods into India for just over ten years, when trade restrictions were eased. However, there are several disadvantages to this method of marketing to rural India.

Without a manufacturing facility in India, a company has no ties to India's already challenging distribution network, thus making sales even more difficult. In addition, Indian consumers tend to feel more loyalty and trust toward locally made products. The aforementioned Thumbs Up and Coca-Cola scenario also illustrates this fact. Though Thumbs Up is a cola of lower quality that Coke, it is a locally made Indian brand that rural consumers can relate with.

Consequently, Coca-Cola is third behind Pepsi and Thumbs Up in the Indian soft drinks market. Partnering with or buying out an existing Indian company, as well as importing from overseas, are all viable ways to get packaged consumer goods into rural India. Based on the past experiences of multinational corporations entering the market, partnering is the most successful option.

Distribution networks in emerging markets tend to be very unique and often times disjointed. India is no exception. Before a multinational corporation even considers entering India's rural market, it is important to first get an understanding of the current distribution system characteristics as well as the ways that the system is likely to change over time. In doing so, a company can assess whether or not accurate and timely product distribution can be achieved without first investing in the distribution networks. Some of the characteristics unique to rural India's distribution networks include the modes of transportation used as well as the point of sale. Despite the challenges of the rural Indian distribution environment, there have been distribution successes from multinational corporations.

Over three million retail outlets in India are reached by companies that produce packaged goods. Methods of transportation used include camels, bulldrawn carts, bicycles, trucks, and trains. In addition, poor roads and unreliable electricity are two additional obstacles common to the distribution networks in rural communities.

Though glass bottles are popular in India, breakage can be a serious problem when the glass is carried over bumpy roads in the back of a truck. Companies must be prepared to design packages for their products that will be capable of withstanding these types of conditions.

The retail establishment where most rural consumers purchase their day-to-day goods is at a kirana or street shop. These small open stalls line the streets and are approximately the size of a living room. Consumers purchase everything from bananas to razors at a kirana. With over 2.5 million kiranas throughout India's rural towns and villages, keeping store shelves stocked is one of the main challenges to consumer goods manufacturers. In order to reach these local shops and establish a brand presence in them, companies need substantial amounts of working capital and a large committed sales force.

In spite of all the distribution challenges, there have been several multinational corporations that have experienced great successes in tackling the distribution networks. Hindustan Lever has been able to build a distribution network in India that directly serves 800,000 stores and uses wholesalers and distributors to reach another 3.5 million outlets. Not only does this help

Hindustan Lever move products from manufacturing facilities to retail outlets, it also provides a large deterrent to potential competitors. In addition to the distribution networks that reach local stores in India, Hindustan Lever began using an experimental concept called Shakti distributors.

They implemented this tactic in 2000 to get their products into some of the most remote rural areas. Douglas Bullis calls this a multilevel marketing system, where independent distributors sell products directly to consumers and earn a commission on the products they sell, plus for other distributors they recruit. It is similar to the way Amway and Mary Kay distribute products in the U.S. Shakti distributors are rural Indian women who partner with Hindustan Lever to receive training in micro-business skills, which includes a Personal Digital Assistant (PDA) to access product prices.

They purchase HLL products at cost and sell them to their villages for a profit. This unique method of distribution gets products beyond the typical reach of HLL's distribution networks. In spite of the unusual modes of transportation and the challenge of supplying small kirana, creative distribution methods have produced success for several multinationals in the rural Indian market.

When approaching the task of designing a package for the rural Indian market, all of the aforementioned factors must be considered. Multinational corporations that have been successful with marketing and packaging consumer products for rural India have taken time to research the target market. They built an insightful and unbiased understanding of the characteristics that make it unique. As a result of this research, two of the most effective elements of a package designed for rural India include the size and visual communication. Material usage is also another important element for the packaging engineer to consider

## Small Incomes

Due to the fact that rural Indians have small disposable incomes and very little storage space, one of the most popular concepts to hit the rural market has been sachets. Sachets are plastic pouches that contain approximately 20 milliliters (.68 oz.) of product.

Sachets were first introduced to India in the 1990's by an Indian company selling a 10-milliliter sachet of Velvette shampoo. Before the sachet, shampoo in India was only available in larger bottles, therefore limiting its sales success among people with small incomes. Sachets meet the needs of the rural consumer in several ways. Sachets are inexpensive, they occupy a small amount of space, and they allow consumers to experiment with new products that they may never have tried before.

Coca-Cola is another company that has found success by thinking small. In a packaging change aimed directly at the rural and lower-income markets, Coca-Cola launched a new 200 mL (6.8 oz.) bottle for the equivalent of 10 cents in 2001. After introducing the smaller size bottle, sales increased 34 percent by the end of the first quarter in 2002. Packaging in smaller units clearly helps to increase the affordability of products for rural Indian consumers.

## Communication

The rural area is a market where large portions of the population are illiterate. So, when packaging consumer products for rural markets, companies must use prominent logo symbols and logo colours to assure that illiterate consumers will be able to recognise the products. Therefore, communicating brand values through the package rather than with words becomes essential. Emotional Surplus Identity (ESI) is a concept that uses the shape, colour, and content of a package to differentiate a brand in the eye of a consumer.

By creating a bond with the consumer through the package, companies are able to establish a relationship that encourages repeat purchases. Loud, bright colours are typically used on packages to differentiate a product from the others on the shelf and to create a lasting impression in a consumer's mind. Another technique used by multinational corporations has been tailoring products, including changing brand names, to give them a rural image. In the eyes of the consumer, branded products are associated with quality and value.

Nirma, the largest selling detergent in the world, found success in the rural Indian market by using unelaborate packaging to position their product as one that cleaned well yet was affordable. While this technique is not the most eye-catching, it allows rural Indian consumers to experience the benefits of a branded product without requiring elaborate or expensive packaging on the part of the multinational corporation. Cost is not only a factor that influences a consumer's decision.

Multinational corporations also address cost when evaluating various packaging options. For example, meeting the needs of consumers by packaging products in small quantities increases the packaging costs for a company in comparison to a large bottle of product. One way companies are able to keep the prices of sachet-type packages down is partially due to lower government duties on small packs. In some instances, it can actually be cheaper for a consumer to purchase sachets rather than a bottle of product. For example, a 100-milliliter (3.4 oz.) bottle of Pantene shampoo retails for 61 rupees whereas 100 milliliters worth of sachets sells for 40 rupees (88 cents). By thinking small, using pronounced colours and logos, and planning for material usage, multinationals can create packages that meet the needs of the rural Indian consumer.

## MEDIA FOR RURAL MARKETING

The concept of rural marketing has to be distinguished from Agricultural marketing. Marketing is the process of identifying and satisfying customers needs and providing them with adequate after sales service. Rural marketing is different from agricultural marketing, which signifies marketing of rural products to the urban consumer or institutional markets.

Rural marketing basically deals with delivering manufactured or processed inputs or services to rural producers, the demand for which is basically a derived outcome. Rural marketing scientists also term it as developmental marketing, as the process of rural marketing involves an urban to rural activity, which in turn is characterised by various peculiarities in terms of nature of market, products and processes.

Rural marketing differs from agricultural or consumer products marketing in terms of the nature of transactions, which includes participants, products, modalities, norms and outcomes. The participants in case of Rural Marketing would also be different they include input manufacturers, dealers, farmers, opinion makers, government agencies and traders.

The existing approach to the rural markets has viewed the markets as a homogeneous one, but in practice, there are significant buyer and user differences across regions as well as within that requires a differential treatment of the marketing problems. These differences could be in terms of the type of farmers, type of crops and other agro-climatic conditions. One has to understand the market norms in agricultural input so as to devise good marketing strategies and to avoid unethical practices, which distort the marketing environment.

Many of the inputs used for production process have implications for food, health and environmental sectors. Rural marketing needs to combine concerns for profit with a concern for the society, besides being titled towards profit. Rural market for agricultural inputs is a case of market pull and not market push. Most of the jobs of marketing and selling is left to the local dealers and retailers. The market for input gets interlocked with other markets like output, consumer goods, money and labour.

The importance of rural marketing can be understood from the fact that today modern inputs i.e. diesel, electricity, fertilisers, pesticides, seeds account for as much as 70% of the total cash costs and 23% of the total costs incurred by the farmers in the Green Revolution areas. Further the percentages were higher at 81% and 38% for small; farmers owning 1.85 hectares of land.

## Strategic View of Rural Marketing

Rural marketing in India is not much developed there are many hindrances in the area of market, product design and positioning, pricing, distribution and promotion. Companies need to understand rural marketing in a broader manner not only to survive and grow in their business, but also a means to the development of the rural economy. One has to have a strategic view of the rural markets so as to know and understand the markets well.

In the context of rural marketing one has to understand the manipulation of marketing mix has to be properly understood in terms of product usage. Product usage is central to price, distribution, promotion, branding, company image and more important farmer economics, thus any strategy in rural marketing should be given due attention and importance by understanding the product usage, all elements of marketing mix can be better organised and managed.

Increasing specialisation in the farming sector has marketers to this strategy. The marketer under this strategy has to design location and carry out farmer specific promotional campaigns. Recommending the use of the products at micro level would result in increasing productivity of the input and thereby increasing the image and the sales of the product can raise the input demand for rural markets.

Joint or co-operative promotion A personalised approach is required under this strategy of rural marketing. Under this approach there is a greater scope for private sector and farmer organisation to get into input supply and especially into retail distribution, as it is a low risk activity. In order to reap the benefits of, the economies of the scale a rural marketer has to resort to bundling of inputs. 'Bundling of Inputs' is the process by which the marketer would provide a bundle of products to the retailer so that he can meet the requirements of the farmers in one place.

The village level co-operatives and other agencies can play an effective role in the distribution of inputs. Establishing linkages with financial agencies and other input sellers can help greatly as the bank credit plays an important role by making the purchase possible. Management of Demand A marketer apart from maintaining good supplies in terms of quality and quantity also has to focus on the demand side of the operations also.

Continuous market research should be undertaken to assess the buyer's needs and problems so that continuous improvements and innovations can be undertaken for a sustainable market performance. Developmental marketing refers to taking up marketing programmes keeping the development objective in mind and using various managerial and other inputs of marketing to achieve these objectives.

A prerequisite for developmental marketing is Development Market Research, which can be termed as the application of marketing research tools and techniques to the problems of development. The research tools of marketing like product testing tests marketing, concept testing and media testing or message test and focus groups are used in this work. Developmental marketing has started to find its roots in India where researchers are using focus groups and products tests to learn more about rural markets and products needs and USPs (Unique Selling Proposition) can be tried out.

Media rural marketing uses both kinds of media i.e. the traditional media as well as the modern media. The traditional media includes puppetry, drama, folk theatre e.g. tamsaha, nautanki, street plays, folk songs, wall paintings and proverbs. Marketer uses traditional media as it is more accessible, personalised, familiar and carries a high potential for change.

The modern media includes the print media, the television and the radio USPs. The companies use different formats to influence the target audience in order to produce the desired results. Extension Services. There are several limitations of rural marketing in the Indian context, this leads to the need for extension services to supplement the efforts of the firms engaged in rural marketing.

The various extension services could include credit facilities, competitions among the farmers, educating the farmers regarding the appropriate agricultural practices, etc. Extension services would thus play a crucial role in the development of rural marketing in India.

## Development of Rural Marketing

Markets have long been a familiar and essential feature of the historical landscape, central places of exchange at which peasants, townspeople, landholders, and rulers have historically converged in India to conduct wholesale and retail trade, to gather news and information, and to engage in various social, cultural, religious, and political activities.

Thus, like their counterparts in other societies, markets can be viewed as microcosms containing a representative array of the elements comprising a regional environment. Markets provide a compressed display of an area's economy, technology, and society—in brief, of the local way of life. To investigate the historiographical rupture provided by the bazaar, let us begin by contextualizing, by drawing its spatial and textual coordinates in relation to the more scrutinized sites in the historiographical landscape—those of caste and village.

Like caste, the village was isolated by the colonial discourse as one of the major sites at which the "real India" was knowable. And like caste, the village—cast in the role of the "Indian village republic"—attained paradigmatic status as a representation of rural society and as a template of the structure and organization of indigenous society and economy.

From such stirrings in the pages of early-nineteenth-century administrative reports, where the idea of the "village republic" or "village community" was "primarily [as] a political society," there emerged a second notion of it as "a body of co-owners of the soil." Subsequently, it developed into the "emblem of traditional economy and polity, a watchword of Indian patriotism."

The political version of village India issued from the pens of Sir Henry Maine and Karl Marx who, although "poles apart in other respects . . . came together retrospectively as the two foremost writers who have drawn the Indian Village Community into the circle of world history. In keeping with contemporary—Victorian—evolutionary ideas and preoccupations, both saw in it a remnant or survival from what Maine called 'the infancy of society.'

"For both men the village denoted a community of interests forged by collective economic and political interests: economic in that the "organized society" of the village tied together by real or fictive kinship held land and cultivated it jointly, political in that authority was wielded by the village council of five, the *panchayat*. Such a view of village India as self-sufficient economically and politically meshed commodiously with the rising nationalist sentiment of the late nineteenth and early twentieth centuries.

Romesh Dutt, for one, deployed these images in attacking the colonial condition whereby such "ancient and self-governing institutions" had passed away "under the too centralised administration of British rulers," their destruction construed as proof that "an alien Government lacks that popular basis, that touch with the people, which Hindu and Mahomedan Governments wisely maintained through centuries."

Periodically transformed but never completely transfigured by shifting intellectual currents, the representations of this imagined village persisted well into the twentieth century, the village's many-layered images accommodating the projections of colonial administrators and scholars, theoreticians, and nationalist leaders and scholars. Although scholars since the 1950s have been steadily dismantling the village that the Raj built—some argue that this edifice has by now not only been condemned but also razed—the mythologized village, like caste, has not entirely relinquished the considerable analytical and theoretical domains it annexed and possessed for more than a century and a half.

Indeed, it has persisted, although with less vigor than has caste, as one of the "few simple theoretical handles [that have] become metonyms and surrogates for the civilization or society as a whole." The concept of the village republic has its roots as much in "European ideas" as in "Indian prejudice." "Like the Mughals, the Nizam of Hyderabad, Sultans of Mysore, and Nawabs of Dacca and Arcot worked revenue systems in which ruling elites knew the countryside only as a set of points for revenue collection.

At every point, settled agriculture and trade generated state revenue." In other words, the colonial state settled on the village as the core of the Indian social, economic, and political body because of its primary interests in maintaining law and order and in extracting taxes from the subject population. Fiscal concerns demanded that the countryside be conceived of as a fixed set of points. An instrumental space for the project of the colonial state, *the* Indian village, as constructed in the colonial discourse, did not, however, fit squarely with British revenue arrangements.

In north India the unit of revenue management was the *mahal* (estate, parcel of land), which increasingly did not coincide with the *mauza* , the revenue village; most settlements, moreover, were made with specific individuals (landholders, or zamindars) and not the village community. The idea of the village republic also ran counter to the *raiyatwari* settlement of south India, which sought to engage the cultivator *(raiyat)* directly. "For, if the village was the basic unit of agrarian organization," as Burton Stein has noted, the "proposal for revenue paid by individual cultivators on specific fields appears misconceived."

Explaining away this contradiction required a rhetorical incorporation that "altered the idea of 'raiyat,' who is changed from being a part of a corporate village body into an equivalent of 'tenant,' thereby generating the transcultural metaphor or analogy—Indian sovereign or East India Company to landlord, raiyat to individualized tenant. In this way both the corporate village and the individual peasant cultivator were preserved . . . while at the same time maintaining the purportedly-direct historical relationship between cultivator 'tenant' and government 'landlord.'

"Contradictions between revenue theory and revenue practice notwithstanding, village India remained at the core of the colonial ideology precisely because it was more valuable as an invented tradition. And in this capacity it serviced colonial power and knowledge, which, to follow the lead of Foucault and others, focused on the manipulation of the body, relying on techniques of discipline and technologies of power to fix people in space by restricting or encouraging their movements and actions and their development and reproduction.

Furthermore, the historically constituted and constructed notion of village India illustrates not only how the technology of colonial power conceptualized one key site—categorizing, classifying, rationalizing, and delimiting in space indigenous society—but also how such ideas expressed the imaginative geography of Orientalism.

The village, in other words, comprised one of the many sites or units (others being criminal tribes, urban spaces, forests, women, and communities of one sort or another) appropriated by colonialism. It therefore speaks to us about the workings of colonial power, not only in exercising political, social, and economic control and domination but also in inscribing itself into the domain of culture and consciousness. A site at which colonial power could produce the "colonized as a fixed reality," the village also qualified as a worthy object of colonial knowledge and power because it was "at once an 'other' and yet entirely knowable and visible."

As an Other, it belonged to a different time; it embodied tradition. The village as tradition, moreover, strengthened the colonial ideology because it represented the backwardness of the subject peoples—it legitimated the right of the rule of modernity, of the Raj. Thus the notion of a conservative rustic population glued to its villages reflected the penchant of disciplinary power for fixing people in space and time, a penchant whose preferred category was naturally a sedentary population.

How closely this tied in-as cause or effect or both-with the changes sweeping the subcontinent over the first century of colonial rule remains to be worked out fully. (So do the many applications of this process, whether targeting other sites or units, e.g., caste, or other groups in the population, e.g., women.) But initial findings suggest that changes in ecology (e.g., extensive deforestation, the advance of the plow) and in modes of agricultural production and exchange, and the shifting fortunes of the once prominent nomadic and pastoral sector, all converged on advancing settled agriculture and peasant petty commodity production.

Stratification within the village community may have occurred as early as the first millennium B.C. The "self-sufficient village," as Romila Thapar states, may have been undermined by the integration of villages via "horizontal links" with "local markets and fairs, networks of religious centres playing an economic role as well and trade in essential items by itinerant herders, artisans and traders." Similarly, Cohn and Marriott note the many connections linking villages to the wider world, ranging from "trading networks" to "networks of marriage ties" to "political networks" consisting "primarily of ties of clan and kinship among rulers and the dominant landlord groups of the countryside."

Morris E. Opler's classic 1956 article identified various "extensions" of the Indian village, fashioned by common origin and descent, with a cluster of villages encompassing an area of seventy square miles; village exogamy, which established ties to other villages; ties of caste to people of similar castes residing in other villages; customary work obligations involving artisans or workers from other villages; supravillage religious or political movements; pilgrimages to sacred sites and visits to the village by religious specialists; the pull of the market town, which offered goods and services not available in the village; and migration to take advantage of education facilities available elsewhere.

"Networks of trade, worship, royal authority, kinship, and caste," as David Ludden has observed recently, "enmeshed a characteristic South Asian village in 1750 within a web of social relations that was essential to agricultural production." These links, networks, and extensions, however, have yet to be mapped out fully. Because the village was elevated so far above all other sites in significance, much of the rural landscape remains beyond the pale of scholarly investigation.

The peripheralization of other sites is especially apparent regarding the "whole subject of agricultural marketing." As Shahid Amin's introduction to the glossary of the colonial ethnographer William Crooke notes, the "prior notion of the changelessness of the physical world of the Indian peasant," led to undue emphasis on the "*production* process" and put "*exchange* relations in parenthesis." Thus, in Crooke's reconstruction, the world of his peasants is "principally inhabited by implements and gadgets, utensils and appliances." "Awkwardly squashed between weights and measures and the rituals of rural life, 'trade and moneylending' are almost pushed out of reckoning."

In part this oversight reflects an historiographical orientation that "has often been peculiarly antagonistic to the rural market; on the basis of rural-romantic or primitive-communistic views of rural self-sufficiency, it has viewed the market as a specifically alien institution. In particular it has often portrayed the denizens of the market as low types who were able to steal the major part of the peasant's produce."

In part the historiographical blinders stem from the "obsessive British concern with Indian Land Revenue." In part the bazaar has remained largely occulted in a historiographical terrain over which the dominant monuments of village and caste have cast their giant shadows. The biases of the historiography, that is, reflect the biases of the colonial record in which village and caste were tropes of an unchanging India. Indeed, relatively little scholarly research has been conducted on the crucial role of markets in articulating the economy and society of an area. Nor have they been systematically analyzed as parts of a wider network of markets.

Instead, studies of marketing in India, as for other areas of the world, are predominantly ethnographies of individual marketplaces and their settings and about certain aspects of market exchange. A few works, primarily by economists, have looked at the efficiency of food grain marketing but only to evaluate their functioning as measured against the models of competitive markets.

There are, however, some significant exceptions: Hagen's scrutiny of the system of colonial education in Patna district in the nineteenth and twentieth centuries within the context of the local society's marketing system; Bayly's examination of the roles of "urban, mercantile and service people" in the towns and bazaars of north India in the period from 1770 to 1870; and Wanmali's synchronic analysis of the lowest level of marketing, periodic markets, in Singhbhum district.

Notable also—although largely synchronic in their focus—are the recent studies of local and regional markets and fairs by geographers and anthropologists that concentrate on the spatial, economic, and anthropological dimensions of patterns of exchange. Studies of markets in India as well as in other parts of the world, while continuing to draw their conceptual sustenance from the central-place theory developed initially in the 1930s by the German economic geographer Walter Christaller, offer a diversity of models of market organization.

But they share a common interest in highlighting markets as nodes in a complex pattern of economic and social exchanges organized hierarchically as well as by such factors as economy, geography, transportation, politics, and administration. However, the tendency in the literature—until recently largely the product of geographers and economists—has been to follow the classical idea of central place to see either how markets organize the geography of retailing or how they serve as collection points for the available goods and services of an area.

An underlying theoretical construct in this literature is the notion of a central place as "a settlement or an aggregation of economic functions that is the hub of a hierarchical system which includes other settlements or communities relating to it on a regular basis; . . . the hub of a region because goods, people, and information flow primarily between it and its less differentiated hinterland." Increasingly, locational and economic analyses of markets—with their emphasis on such geographical and economic variables as population, ecology, transportation, and a competitive market economy—have been enriched by studies recognizing other factors relating to social, political, and cultural conditions and circumstances also involved in structuring marketing systems. This body of scholarship, much of it largely synchronic in focus, relies on and reinterprets and modifies central-place theory.

The Indian rural market with its vast size and demand base offers a huge opportunity that MNCs cannot afford to ignore. With 128 million households, the rural population is nearly three times the urban. As a result of the growing affluence, fuelled by good monsoons and the increase in agricultural output to 200 million tonnes from 176 million tonnes in 1991, rural India has a large consuming class with 41 per cent of India's middle-class and 58 per cent of the total disposable income.

The importance of the rural market for some FMCG and durable marketers is underlined by the fact that the rural market accounts for close to 70 per cent of toilet-soap users and 38 per cent of all two-wheeler purchased. The rural market accounts for half the total market for TV sets, fans, pressure cookers, bicycles, washing soap, blades, tea, salt and toothpowder, What is more, the rural market for FMCG products is growing much faster than the urban counterpart.

## The 4A Approach

The rural market may be alluring but it is not without its problems: Low per capita disposable incomes that is half the urban disposable income; large number of daily wage earners, acute dependence on the vagaries of the monsoon; seasonal consumption linked to harvests and festivals and special occasions; poor roads; power problems; and inaccessibility to conventional advertising media. However, the rural consumer is not unlike his urban counterpart in many ways. The more daring

MNCs are meeting the consequent challenges of availability, affordability, acceptability and awareness (the so-called 4 As).

*Affordability*

The second challenge is to ensure affordability of the product or service. With low disposable incomes, products need to be affordable to the rural consumer, most of whom are on daily wages. Some companies have addressed the affordability problem by introducing small unit packs. Godrej recently introduced three brands of Cinthol, Fair Glow and Godrej in 50-gm packs, priced at Rs 4-5 meant specifically for Madhya Pradesh, Bihar and Uttar Pradesh — the so-called 'Bimaru' States.

Hindustan Lever, among the first MNCs to realise the potential of India's rural market, has launched a variant of its largest selling soap brand, Lifebuoy at Rs 2 for 50 gm. The move is mainly targeted at the rural market. Coca-Cola has addressed the affordability issue by introducing the returnable 200-ml glass bottle priced at Rs 5. The initiative has paid off: Eighty per cent of new drinkers now come from the rural markets. Coca-Cola has also introduced Sunfill, a powdered soft-drink concentrate. The instant and ready-to-mix Sunfill is available in a single-serve sachet of 25 gm priced at Rs 2 and mutiserve sachet of 200 gm priced at Rs 15.

*Acceptability*

The third challenge is to gain acceptability for the product or service. Therefore, there is a need to offer products that suit the rural market. One company which has reaped rich dividends by doing so is LG Electronics. In 1998, it developed a customised TV for the rural market and christened it Sampoorna. It was a runway hit selling 100,000 sets in the very first year. Because of the lack of electricity and refrigerators in the rural areas,

Coca-Cola provides low-cost ice boxes — a tin box for new outlets and thermocol box for seasonal outlets.

The insurance companies that have tailor-made products for the rural market have performed well. HDFC Standard LIFE topped private insurers by selling policies worth Rs 3.5 crore in total premia. The company tied up with non-governmental organisations and offered reasonably-priced policies in the nature of group insurance covers. With large parts of rural India inaccessible to conventional advertising media — only 41 per cent rural households have access to TV — building awareness is another challenge.

Fortunately, however, the rural consumer has the same likes as the urban consumer — movies and music — and for both the urban and rural consumer, the family is the key unit of identity. However, the rural consumer expressions differ from his urban counterpart. Outing for the former is confined to local fairs and festivals and TV viewing is confined to the state-owned Doordarshan. Consumption of branded products is treated as a special treat or indulgence.

Hindustan Lever relies heavily on its own company-organised media. These are promotional events organised by stockists. Godrej Consumer Products, which is trying to push its soap brands into the interior areas, uses radio to reach the local people in their language. Coca-Cola uses a combination of TV, cinema and radio to reach 53.6 per cent of rural households. It doubled its spend on advertising on Doordarshan, which alone reached 41 per cent of rural households.

It has also used banners, posters and tapped all the local forms of entertainment. Since price is a key issue in the rural areas, Coca-Cola advertising stressed its 'magical' price point of Rs 5 per bottle in all media.LG Electronics uses vans and road shows to reach rural customers. The company uses local language advertising.

Philips India uses wall writing and radio advertising to drive its growth in rural areas.

## Enabling ICT for Rural India

Over the past decade, India has become the worlds test bed for innovations in information and communication technologies (ICT) serving the rural user. Various reasons explain this emergence.

The most obvious is the search for a solution to a long-intractable problem: that rural India's institutional infrastructure is woefully inadequate. The hope that ICT will help to manage rural India's social, political and administrative challenges and become a viable technology for the provision of health, education and other social services is thus ICT's strongest calling card. An additional expectation is that ICT will improve access to the large underserved market that rural India's 700 million people represent. Both expectations' salience increases with the argument that India has the resources to build an ICT infrastructure, i.e., its large, skilled, cost-efficient IT workforce.

Over three months, from September to December 2004, our team from Stanford University and the National Informatics Center visited nine ICT projects, chosen for diversity and importance. We concluded that all the projects are still experimental. None has yet had a widespread socio-economic impact or even developed a catalytic, replicable approach. Under present operating conditions, financial self-sufficiency is, generally speaking, unlikely. The hope that ICT can build rural capacity for self-development has not been realized.

Some findings were as expected, such as poor infrastructure, high deployment and maintenance costs of the ICT infrastructure and inadequate content for eGovernance. A less expected finding is that

EGovernance services are overwhelmingly the most needed but the least provided services – instead, most projects provide informational services, i.e., generic, non-customized services such as agricultural practices, weather forecasts and contact information; and, secondarily, they provide transactional services, i.e., the exchange of specific, customized informational services or funds between two or more parties, such as email or e-commerce.

The importance of eGovernance arises from the prolonged absence of self-sufficiency in rural areas, which has created an encompassing dependency of rural residents on locally elected officials and bureaucrats. The state needs to undertake two steps to help. First, it needs to parse the general rubric of eGovernance into components based on type, such as: (a) Generic information services about government projects and employment opportunities. (b) Customized information such as land records and birth certificates. (c) Approvals, such as for 'below poverty line' status, and grievance redress. (d) Social services: health, education, entitlement and other social services. (e) Mandatory services: taxation, updating land and population databases and (f) Exchange services: postal, banking and utility services

Second, for each type, the state needs a model for public-private partnerships in the creation and delivery of eGovernance services. Current approaches focus on using the private sector to bypass local officialdom, such as for delivering complaints to district officials. This is an arena most resistant to change. Instead, focusing on automation and back-end digitization of informational and exchange services, and outsourcing digitized social services to the private sector are some viable possibilities.

A second unexpected finding was that high deployment costs of an Internet infrastructure prevent projects from having enough resources to develop content. In addition, the weak (and costly) communications infrastructure has forced projects to store content locally than over the Internet.

These problems are because of the single-user model, i.e., the village kiosk is the only connection from the village. In no country, however developed, can a single user afford to pay for the entire transmission infrastructure. While this implies that public funds such as USO funds are needed, nevertheless a solution to using public funds needs to be carefully designed so as to retain incentives for competition and coverage within the public sector.

# 9

# Rural Development and Poverty Alleviation

Rural development implies both the economic betterment of people as well as greater social transformation. Increased participation of people in the rural development process, decentralisation of planning, better enforcement of land reforms and greater access to credit and inputs go a long way in providing the rural people with better prospects for economic development. Improvements in health, education, drinking water, energy supply, sanitation and housing coupled with attitudinal changes also facilitate their social development. Rural poverty is inextricably linked with low rural productivity and unemployment, including underemployment.

Hence it is imperative to improve productivity and increase employment in rural areas. Moreover, more employment needs to be generated at higher levels of productivity in order to generate higher output. Employment at miserably low levels of productivity and incomes is already a problem of far greater magnitude than unemployment as such. It is estimated that in 1987-88 the rate of unemployment was only 3 per cent and inclusive of the underemployed, it was around 5 per cent.

As per the currently used methodology in the Planning Commission, poverty for the same year was estimated to be 30 per cent. This demonstrates that even though a large proportion of the rural population was "working" it was difficult for them to eke out a living even at subsistence levels from it. It is true that there has been a considerable decline in the incidence of rural poverty over time. In terms of absolute. numbers of poor, the decline has been much less.

While this can be attributed to the demographic factor, the fact remains that after 40 years of planned development about 200 million are still poor in rural India. In 1987-88, the rural poverty line in terms of per capita monthly expenditure was Rs. 131.80. The average incidence of rural poverty conceals wide inter-state differences which suggests that greater attention needs to be paid to the regions which have a greater concentration of the rural poor.

In recent years, several issues have been raised about the methodology of poverty estimation, both by professionals and State Governments. An Expert Group appointed by the Planning Commission is looking into these issues relating to the definition and measurement of poverty. The decline in rural poverty is attributable both to the growth factor and to the special employment programmes launched by the Government in order to generate more incomes in the rural areas. Hence, in its more limited interpretation, rural development has been confined to a direct attack on poverty through special employment programmes, area development programmes and land reforms.

Under the Integrated Rural Development Programme (IRDP) those living below the defined poverty line in rural areas are identified and given assistance for acquisition of product live assets or appropriate skills for self-employment, which in turn, should generate enough income to enable the beneficiaries to rise above the poverty line.

This scheme was launched in the Sixth Plan. Its assessment at the end of the Sixth Plan period revealed several shortcomings. Keeping this in view and the feed-back received from the State Governments, suitable changes were introduced in the guidelines for the IRDP in the Seventh Plan.

The poverty line was based at Rs.6400, but those eligible for assistance under the IRDP had to have an average annual income of Rs.4800 or less. It was assumed that those households with income levels between Rs.4800 and Rs.6400 would be able to rise above the poverty line in the process of growth itself. It was targetted that 20 million families would be assisted under IRDP during the Seventh Plan of which 10 million were new households and 10 million old beneficiaries who had been unable to cross the poverty line and required a second dose. During the Seventh Plan, me subsidy expenditure on IRDP was Rs.3316 crores which was in excess of the target of Rs.3000 crores.

The total investment including the institutional credit amounted to Rs.8688 crores. In quantitative terms, the physical achievement of about 18 million households fell short of the original target of 20 million households but exceeded the cumulative target which was only 16 million families. The sectoral composition indicates that, of all the schemes selected under IRDP, 44 per cent were in the primary sector, 18.5 per cent in the secondary sector and 37.5 per cent in the tertiary sector.

A system of concurrent evaluation of the IRDP programme was also introduced under which data were collected by independent research institutions for the entire country on a sample basis. Statewise details of performance are available. The findings suggest that the IRDP was quite successful in terms of providing incremental income to poor families. However, the number of households able to cross the poverty line was relatively small. It may be partly due to the low levels of initial investment. On the other hand, it is also difficult to expect banks to raise the per capita loan assistance to beneficiaries, given the excessive overdues pending. In order to enhance the economic returns from an asset, it is necessary to integrate this scheme with the development plans of an area so that select activities become viable.

## Training of Rural Youth for Self Employment (TRYSEM)

TRYSEM was introduced in 1979 to provide technical skills and to upgrade the traditional skills of rural youth belonging to families below the poverty line. Its aim was to enable the rural youth to take up self- employment ventures in different spheres across sectors by giving them assistance under IRDP. Later, in 1987 the scope of the programme was enlarged to include wage employment also for the trained beneficiaries.

During the Seventh Plan about 10 lakh youth were trained under TRYSEM, of which 47 per cent took up self- employment and 12 per cent wage employment. The remaining 41 per cent could not avail of either. On the other hand, a sizeable proportion of IRDP beneficiaries who needed training could not receive it. In fact, only 6 to 7 per cent of IRDP beneficiaries were trained under TRYSEM. During 1990-91 the number of youth trained were 2.6 lakhs, of which 70 per cent got employed.

## Development of Women and Children in Rural Areas (DWCRA)

In 1982-83 an exclusive scheme for women was launched in the IRDP, as a pilot project, in 50 districts. In the Seventh Plan it was extended to more districts and at the end of the Seventh Plan period it was in operation in 161 districts. Under DWCRA, a group of women are granted assistance to take up viable economic activities with Rs. 15,000 as a one-time grant to be used as a revolving fund. In the Seventh Plan about 28,000 groups could be formed against the target of 35,000 with a membership of 4.6 lakh women. During 1990-91, against a target of 7,500 groups, 7,139 were actually formed.

While, in principle, this scheme is a sound one, in operationalising it the impadt has been inadequate. This is perhaps due to 'a lack of cohesion among women groups formed under DWCRA and their inability to identify activities that could generate sustained incomes. In this sphere, the role of voluntary organisations would be crucial in organising women to take up group-based economic activities which are viable within the context of an area development plan. Experiments in some States to form women's thrift and credit societies first, and then start them on economic work have been successful.

## Wage Employment Programmes

In 1989, the erstwhile National Rural Employment Programme (NREP) and the Rural Landless Employment Guarantee Programme (RLEGP) was merged into a single rural wage-employment programme called the Jawahar Rozgar Yojana. However, given that in the first four years of the Seventh Plan, the NREP and RLEGP were in operation, a brief review of these two programmes is given below.

## National Rural Employment Programme (NREP)

The entitlement of each State to the Central fund was based on the incidence of poverty and the population of agricultural labourers, marginal farmers and marginal workers with 50 per cent weightage to each. However, the Centre and State shared the expenditure equally on a 50:50 basis. Some broad indicators of the performance both physical and financial are set-out in the table below:

A concurrent evaluation of NREP re vealed that several types of assets were created. with 24.6 per cent expenditure on rural ruatk and 19.1 per cent on social forestry. Construction was a main activity with 11.9 per cent or; schools, 12.1 per cent on houses and 6.4 per cent on panchayat ghars; 6.5 per cent was directed to minor irrigation and 3.3 per cent to wells for drinking water.

## Rural Landless Employment Guarantee Programme (RLEGP)

This was a totally Centrally financed programme introduced in 1983. While most of the objectives and stipulations under this were similar to those of NREP, it was to be limited only to the landless, with guaranteed employment of 100 days. Moreover, there was earmarking of funds specifically for certain activities- 25 per cent for social forestry, 10 per cent for works benefitting only the Scheduled Castes/Scheduled Tribes and 20 per cent for housing under Indira Awaas Yojana.

In the Seventh Plan, Rs.2412 crores were spent and 115 crore mandays were generated with an average expenditure of Rs.21.00 per manday. Only 16 per cent had been spent on social forestry but 22 per cent had been spent on housing,- with over 5 lakh houses created for SC/ST and freed bonded labourers. Rural roads accounted for 22 per cent while other construction, minor irrigation, soil conservation etc. each had a small share.

In the last year of the Seventh Plan, Jawahar Rozgar Yojana (JRY) JRY was laúnched with a total allocation of Rs. 2600 crores to generate 931 million mandays of employment. The primary objective of the programme is generation of additional employment on productive works which would either be of sustained benefit to the poor' or contribute to the creation of rural infrastructure. Under this programme, Centre's contribution is 80 per cent, and 20 per cent is the State's share. The JRY is implemented in all villages in the country.

Central assistance is provided to the States on the basis of proportion of the rural poor in a State/UT to the total rural poor in the country. From the States to the districts, the allocations are made on an index of backwardness which is formulated on the following basis:

– 20 per cent weightage for the proportion of agricultural labourers in the total workers in the rural areas.
– 60 per cent weightage to the proportion of rural scheduled castes and tribes population in relation to the total rural population; and
– 20 per cent weightage to the inverse of agricultural productivity.

Of the total allocations at the State level 6 per cent of the total resources are earmarked for housing under the Indira Awaas Yojana (IAY) which are allotted to the scheduled castes and scheduled tribes and freed bonded labour. In addition, 20 per cent are earmarked for Million Wells Scheme (MWS). In fact, this scheme was launched as a special feature both under NREP and RLEGP in 1988-89.

The objective is to provide open wells, free of cost, to poor SC/ST farmers in the category of small and marginal farmers, and to free bonded labourers. However, where such wells are not feasible, the amounts allotted may be utilised for other schemes of minor irrigation like irrigation tanks, water harvesting structures and also for development of lands of SCs/STs and freed bonded labourers including ceiling surplus and bhoodan lands. A maximum of 2 per cent of JRY funds are to be spent as administrative costs inclusive of any additional staff.

After providing for the above earmarking, 20 per cent of the remaining funds are retained at the district level and 80 per cent are allocated to village panchayats by giving 60 per cent weightage to SC/ST population and 40 per cent to the total population of the village panchayat. The responsibility of implementation of JRY in respect of district share of funds is that of DRDA/Zilla Parishad, but at the village level it is that of the Gram Panchayat.

In case two or more districts/gram panchayats decide to pool the resources together to take up a work for the common benefit of the concerned dis-trict/ panchayat, the arrangement is permissible. Works can be taken up for execution during any part of the year whenever the need for generating supplementary employment is felt, preferably during the lean agricultural season but could continue during the busy agricultural period too, if required.

A maximum of 10 per cent of the annual allocation can be used for incurring expenditure on maintenance of such assets at the district/gram panchayat levels which have been created under the erstwhile programme of NREP/RLEGP or have been created under JRY and have not to be taken over by a department of the State Government. There is earmarking of resources at the district level too but after deducting for administrative and maintenance costs, the funds available for sectoral works are untied.

However, there is no sectoral earmarking of resources at the village panchayat level except that 15 per cent of the annual allocation must be spent on works directly beneficial to SCs/STs. The types of works, to be taken up under the village panchayats are to be based on the felt needs of the people. There js 30 per cent reservation for women. Sixty per cent of the total unit cost is to be spent on wages and 40 per cent on materials. Contractors and other intermediaries are not permitted in the execution of the works.

Minimum wages, fixed by the State, are required to be paid. There are considerable inter- State variations in the minimum wage - rates, ranging from Rs. 13.70 to Rs.34.00 per day for unskilled work. These differences account for variations in the Statwise unit cost of generating one man-day of employment ranging from Rs.22.83 to Rs.56.67. While in the earlier wage-employment programmes, part of the wage payment had to be in kind, in terms of certain quantity of foodgrains, under the JRY this was made optional.

Consequently, while in 1986-87 the offtake of foodgrains was as high as 22 lakh tonnes, in 1990-91 it was only 1.36 lakh tonnes. Road construction was the primary activity accounting for a little less than 30 per cent of the expenditure while minor irrigation, housing, construction of school and community buildings, wells and social forestry were the other sectors where JRY funds flowed.

The expenditure on housing and wells increased between 1989-90 and 1990-91, while that under social forestry declined. The earmarking for housing under Indira Awaas Yojana and for wells and the MWS would definitedly have contributed to this. Under the IAY, 8.6 lakh houses have been constructed and 2.6 lakh wells have been dug under MWS. These would have benefitted the scheduled castes and the scheduled tribes.

A concurrent evaluation has been initiated which will be completed by the end of the year. In the meantime, a quick evaluation of JRY has also been undertaken.

## State level Employment Programmes

### Maharashtra Employment Guarantee Scheme (EGS)

The Maharashtra Employment Guarantee Scheme (EGS) is a unique experiment which was started in 1971-72 for providing gainful employment in rural areas and "C" class muncipal areas. Guaranteed unskilled manual work is provided to adults who register themselves for work. Only productive works with unskilled wage component of more than 60% are taken up under the scheme. In the last two to three years, the EGS has been improved and modified.

Under the "Shram Shakti Divare Gram Vikas", individual beneficiary's scheme will be taken up at the cost of the Government in the case of lands owned by small and marginal farmers, but for other categories 50 per cent of the expenditure will be borne by the concerned cultivator/beneficiary. Again, a horticulture programme with the target of covering a total of 10 lakh ha. during the Eighth Plan has been launched at Government cost on lands of SC/STs/small farmers/NTS. On other lands, Government and the beneficiaries bear the expenditure on materials in the ratio 75:25.

The resources for the scheme are raised by the State Government by (i) levying a number of taxes/additional taxes/surcharges on profession, trade, motor vehicles, sales tax, irrigated agricultural land, land revenue and non-residential land and (ii) a contribution equal to the net collection of these levies made by the State Government.

The expenditure on me EGS has varied over the last six years from about Rs.288 crores in 1987-88 to an anticipated expenditure of Rs.200 crores in 1991-92 and employment generation from 18.95 crore mandays to an anticipated level of 7.50 crore mandays. Wages paid under the scheme are not lower than the minimum wages for unskilled agricultural labour.

The scheme has resulted in a significant reduction in the incidence of unemployment in rural areas. Average daily unemployment rates in rural Maharashtra have declined from 7.20% in 1977-78 to 3.17% in 1987-88. It would also have contributed to some extent towards the decline in rural poverty from 60.4 per cent in 1977-78 to 36.7 per cent in 1987.88. The scheme has also helped in keeping an upward pressure on wages in rural areas. The EGS has benefitted a large number of women too, with nearly 60 per cent of the workers on EGS sites being women.

In view of the significant positive impact of this scheme on employment, earnings and levels of living of rural people in Maharashtra, the experiment could well be a model for similar schemes in other States.

### Special Employment Programme of Gujarat

A special employment programme was introduced in Gujarat in 1991 under which two districts, Dang and Gandhinagar, were selected for achievement of zero unemployment and, in the remaining districts, additional employment opportunities will be created in the rural areas. A plan is being worked out with the objective of providing self-employment to those below poverty line, as well as opportunities for wage employment for those who seek it.

The DPAP was launched in 1973 in arid and semi-arid areas with poor natural resource endowments. The objective was to promote more productive dryland agriculture by better soil and moisture conservation, more scientific use of water resources, afforestation, and livestock development through development of fodder and pasture resource, and in the long run to restore the ecological balance. The DPAP covers 615 blocks of 91 districts in 13 states.

Under DPAP, funds are allocated on the basis of the number of blocks covered under the programme in each district at the rate of Rs. 15 lakhs per block, having a geographical area upto 500 sq. kms., Rs. 16.5 lakhs per block with an area between 500-1000 sq. kms., and Rs.18.5 lakhs for blocks with an area exceeding 1000 sq. kms. The allocations are shared between the Centre and the States on a 50:50 basis. The financial and physical achievements under the programme are given below.

The Programme Evaluation Organisation of the Planning Commission has been entrusted with the task of evaluating the DPAP. These programmes have been running for many years and there is no evidence that drought -proofing has been achieved in any of the DPAP blocks. Yet there are cases where voluntary effort has succeeded in achieving this objective at a micro-level. A more concerted and coordinated effort would be required with greater use of scientific data, detailed working of cost norms for different activities and efficient planning along micro watershed lines. Emphasis is laid on training of project staff at the district/watershed level for preparation of plans and creating awareness among the people of the project areas.

Stress is also laid on the need for developing effective liaison between agricultural research agencies and implementing agencies for effective transfer of technology. To ensure participation of people in planning and implementation of the programme, various measures have been taken, such as preparation of watershed development plan with the help of the people in the watershed itself under the guidance of technical experts, and adequate local representation in the Watershed Development Committee set up for implementation of the project. In the Eighth Plan renewed thrust along these lines will be given to the DPAP.

## Land Reforms

The land reforms policy has consisted of the following:

- Abolition of intermediaries;
- Tenancy reforms with security to actual cultivators;
- Redistribution of surplus ceiling land;
- Consolidation of holdings; and
- Updating of land records.

In the first stage of the programme there was, in the early fifties, the abolition of 'Zamindari', which covered 40% of the land area of the country benefitting 20 million cultivators. Fifteen lakh areas of wasteland were also vested in the State. In the process of implementing this measure, old Zamindars succeeded in retaining large tracts for self-cultivation.

There are tenancy laws in all the States except Nagaland, Meghalaya and Mizoram. They provide for conferment of ownership on the tenant by the State, acquisition of ownership by tenants on payment of reasonable compensation, security of tenure and fixation of fair rent. Certain categories such as widows, members of armed forces, minors, etc. are treated specially under these laws. In certain other cases, provision is also made for limited right of resumption. However, the implementation of these laws in States has been quite varied. West Bengal, Karnataka and Kerala have achieved more success than the other States.

In West Bengal, 14 lakh share- croppers have been recorded under the 'Operation Barga'. Karnataka set up land tribunals to settle tenancy issues and these decided in favour of 3,00,000 tenants involving 11 lakh acres of land. In Kerala, through the tenants' association, applications of 24 lakh tenants for conferment of ownership were accepted.

However, on the whole, tenancy reforms have not achieved the desired results as the incidence of informal oral or concealed tenancies is very high. In fact, it was envisaged in the Sixth Plan that legislative measures to confer ownership rights to tenants would be introduced in all States by 1981-82. This is still an issue that has to be tackled.

Of the 72.2 lakhs acres of land declared surplus, 46.5 lakh acres had been distributed by the end of the Seventh Plan and 25.7 lakh acres are still to be distributed.Consolidation of holdings has made progress in some States while in others it is yet to make a beginning. Fifteen States have passed laws for consolidation of holdings. Those not having laws are Andhra Pradesh (in select areas ofAndhra Pradesh), Tamil Nadu, Kerala, Pon-dichery and the North-Eastern States. Tenants, share croppers and small landowners, have a fear that consolidation favours the larger farmers. So far, about 1494 lakh acres have been covered.

The allotees of surplus ceiling land require assured access to inputs. This is being done under a scheme wherein Rs.2500/- per hectare are provided for land development, purchase of inputs and for meeting other needs. States have been asked to make 40% allotment of surplus land to women and the remaining in joint names of husbands and wives.

The Centrally Sponsored Scheme for the strengthening of revenue administration and updating of land records was introduced during the Seventh Plan. Under this scheme, by the end of the Seventh Plan Rs.25.57 crores have been allocated to 29 States and Union Territories for purchase of equipment and strengthening of training infrastructure.

Nineteen pilot projects for computerisation of land records have been taken up, one in each major State. These are fully financed by the Central government at the rate of Rs.25 lakhs each. The project envisages computerisation of the record of rights in the first stage, and both input and output will be in the local language. Computers are being installed at the Tehsil and district headquarters, with the objective of online updating and making available a copy of the record of rights to cultivators on demand. The project is nearing completion in Morena district of Madhya Pradesh and Dungerpur of Rajast-han. The progress in other States needs to be expedited.

## Development Administration

During the Seventh Plan, the various rural development programmes were planned and implemented by a single agency at the district level called the District Rural Development Agency (DRDA). However, at the block level there was an attempt to return to the earlier community development pattern. But this was not easy, as the BDO had lost effective control over the Extension Officers who were functioning under their own departmental heirarchies.

Also, there had been a tremendous increase in the volume of work and in the funds flowing at the block level. As against Rs. 17 lakh per year in the sixties, it became Rs. 1 crore per block per year. This put an enormous burden on the administrative system. A Committee was set up to review the existing administrative arrangements for rural development, which submitted its report in 1985. It reemphasised the need for decentralised planning at the district level and below. It opined that where Zila Parishads were in existence rural development programmes should be transferred to them.

This would ensure participation of local representatives in planning and they in turn would reflect the needs and aspirations of the local people. Of course, they would also be accoutable to the people they represent. In States where Zila Parishads are not in existence, the setting up of District Development Councils with Government officers as the Chief Executives was suggested.

In either case, it was envisaged that planning and implementation of sectoral activities would be decentralised and integrated into a unified activity, with horizontal coordination at the district level. Similarly, at the block level too, an integrated area plan was imperative, based on availability of local skills and resources. However, no uniform pattern was adopted across States. In 1989-90, the introduction of the Jawahar Rozgar Yojana, wherein it was stipulated that the funds would be placed at the disposal of the village panchayats, marked a shift towards democratic decentralisation, with certain funds and powers vested in the gram panchayats for development.

## Panchayati Raj

Panchayati Raj Institutions are in existence in almost all the States and UTs but with considerable variations in their structure, mode of election, etc. In 14 States/UTs, the three-tier system exists, while four States have tw6;tier and nine states/UTs have one-tier system. In Nagaland, Arunachal Pradesh, Meghalaya, Mizoram, a large part of Manipur and some other hilly areas of North-Eastern States, these institutions are established in accordance with the traditions and customs of the village.

At the end of the Seventh Plan, there were 2,17,300 Gram Panchayats, 4525 block Panchayati Samities and 330 Zila Parishads in the country. The tenure of the elected bodies is between 3 and 5 years. However, Panchayati Raj Institutions suffer from inadequate resources, both financial and technical. In most of the States, they are not entrusted with enough powers and financial responsibilities. With a view to strengthening the Panchayati Raj Institutions and making them a vibrant instrument of local self-Government, a process of grassroot level consultation was initiated towards the end of the Seventh Plan period.

For the first time, Panchayati Raj Sammalens were held in the different regions during 1989 wherein delegates comprising Sarpanches, Taluka/Block Panchayati Samities President, Chairman of Zila Parishad and Chariman of Muncipal Committees/Town Area Committees and Notified Area Committees participated. The main objective is to make these institutions strong, reflecting the felt needs of the people.

To revitalise the Panchayats, a Constitution Amendment Bill (Constitution 72nd Amendment Bill, 1991) was introduced in Parliament in 1991. The Constitution Amendment Bill itself provides for, inter-alia, a 'Gram Sab-ha' in each village, constitution ofpanchayats at village and other level or levels, direct elections in all States to Panchayats at the village level and intermediate levels, reservation for scheduled castes and the scheduled tribes in proportion to their population and reservation of not less than one-third of the seats for women, fixing tenure of five years for local authorities, and holding elections within a period of six months in the event of supersession of any such authority.

The State legislatures are required to devolve powers and responsibilities on the panchayats for preparation of plans for economic development and social justice and for implementation of development schemes. Grants-in-aid to panchayats from consolidated fund of the State as also conferment of powers for levy of taxes, duties, tolls and fees are provided for.

Further, it envisaged the setting up of a Finance Commission within one year of the Amendment Bill and, thereafter, every five years to review the financial position of local authorities. While the Bill has been introduced in Parliament, it is yet to be debated and passed. Once enacted, democratic decentralisation will be achieved through the Panchayati Raj Institutions.

## VOLUNTARY ACTION

Recognising the important role of voluntary agencies' in accelerating the process of social and economic development, the Seventh Plan placed a great deal of emphasis on people's participation and voluntary action in rural development. The role of voluntary agencies has been defined as providing a basis for innovation with new approaches towards integrated development, ensuring feed-back regarding impact of various programmes and securing the involvement of local communities, particularly, those below the poverty line. The need for a cadre of trained animators and social organisers was recognised and a massive programme for training the identified persons was prepared with the help of establishing Voluntatry Organisations.

Further, the scheme of organisation of beneficiaries of anti-poverty programmes which was undertaken on a pilot -basis for two years from 1986-87 was continued during the Seventh Plan period. This scheme was intended to increase the awareness and strengthen the bargaining position of the beneficiaries of anti-poverty programmes so as to help them get the maximum benefits from the programmes meant for their economic uplift.

This was to be done through awareness generation camps, which were organised with the assistance of voluntary organisations. At the Central level, the Council for Advancement of People's Action and Rural Technology (CAPART) is the agency for providing and assisting voluntary action in the area of rural development. Its funds comprises mainly grants from the Government of India.

Programmes of the Ministry of Rural Development including, IRDP, JRY, DWCRA, TRYSEM, Organisation of beneficiaries, Accelerated rural water supply. Central rural sanitation programme etc. are implemented by. voluntary agencies through the assistance of CAPART. In addition, CAPART has takeifthe initiatives in promoting a variety of activities for transfer of technology, people's participation, development of markets for products of rural enterprises and promotion of other developmental activities and delivery systems in the nongovernment sector.

## Special Employment Programmes

Elimination of poverty continues to be a major concern of development planning. Expansion of employment opportunities, augmentation of productivity and income levels of both the underemployed and employed poor would be the main instrument for achieving this objective during the Eighth Plan. However, even an employment oriented growth strategy will achieve this goal only in the medium and long-term.

In the meantime, short-term employment will have to be provided to the unemployed and underemployed, particularly among the poor and vulnerable sections, through the existing special employment programmes namely the IRDP and JRY. However, it must be recognised that while they meet the short-term objective of providing temporary work to the unemployed they must contribute to the creation of productive capacity of areas and/or individuals.

This would be better achieved by a greater integration of the existing special employment programmes with other sectoral development programmes, which, in turn, would generate larger and more sustainable employment. Given the enhanced outlay for\rural development' in the Eighth Plan, it is necessary that resources are utilised for building up of rural infrastructure, which is an essential pre-requisite for a more sustained employment and development.

All weather roads need to be given priority, particularly in tribal, hill and desert areas, where inaccessibility to markets and to information and input is a severe bottle-neck. Minor irrigation works and water harvesting structures are vital in order to conserve the scarce water and schemes for soil conservation and social forestry would go a long way in reducing soil erosion and top soil water cup - off as well as wherever required school buildings and primary health centres and sub-centres need to be constructed.

The demand for these would vary between regions and even districts. Hence, a certain degree of flexibility would have to be built into the programme, leaving the choice to the people at the local level based on their needs and priorities. In addition, the planning and implementation of the rural development programmes must enable greater self-help by the people and their participation in programmes through panchayati raj institutions, cooperatives and other self-managed institutions.

This will mark a reduction in the dependence on the present development administration for delivery. However, this should not be interpreted as a greater move towards 'privatisation' or leaving the rural poor to look after themselves. State intervention will have to continue, in fact, on an expanded scale so as to protect the poor and vulnerable sections from some of the burdens of structural adjustment.

Certain changes would be required in the broad strategy for rural development during the Eighth Plan.

Experience indicates that while poverty alleviation programmes have been successful in providing a certain quantum of employment to people and have led to the creation of some durable assets in the village, there is a perception that the achievements have not been commensurate with the resources spent on them.

Under the IRDP, the very fact that about half the number of beneficiaries have overdues raises doubts about their ability to come out of the debt syndrome. This, it is argued, is due to a low level of assistance which does not generate enough income to repay the loan and for subsistence. However, banks are reluctant to raise the credit limit because of scepticism regarding the repayment capacity of the target groups. It is estimated that about one-third of them do not even have the original asset that was given to them.

The beneficiaries may be forced to sell the asset as they require the money. What is more, even those who have generated sufficient additional income to cross the poverty line may relapse into the category of poor, with additions to the family, loss of assets and non-viability of the activity chosen by him. Similarly, under the Jawahar Rozgajr Yojana, some employment is provided in the lean season and the supplementary incomes thus generated are critical for the survival of many poor families.

But the wages earned under JRY are a very small proportion of the amount required to help him to cross the poverty line. Moreover, while some productive assets are created, which add to the infrastructural facilities available in a village their quality could be improved. Also, often they do not reflect the priorities of the local people but of the panchayat functionaries and their maintenance is lacking.

Under the JRY any additional allocation must be linked to certain backward districts/ blocks, with an element of guarantee of at least 90-100 days of employment per person as under the Maharashtra Employment Guarantee Programme. Only then will it provide the s safety net for the poor unemployed who may find it difficult to subsist in lean seasons. Providing employment of only 15-25 days per person is grossly inadequate.

There is no doubt that a wage-employment programme like the JRY requires to be better targetted. A quick evaluation of the JRY conducted by the Programme Evaluation Organisation supports these observations. The survey shows that on an average, about 15 days of employment was generated per person in 1990-91. At an average wage-rate of Rs. 20.00 to Rs. 25.00, this would yield a supplementary income per person of about Rs. 300 -400 per anum: This is rather meagre in the context of the poverty line of Rs.6,400 during the Seventh Plan, which has been revised upwards for the Eighth Plan.

Also, wage- material ratio of 60 : 40 was not sustainable, as the rising material costs meant more capital for creating durable assets. However, the beneficiaries were happy with the assets created, though their maintenance was somewhat lacking. Also, certain relaxation and changes in the stipulation and guidelines incorporated both in the IRDP and JRY would be required to make them more effective. Under the JRY a certain degree of flexibility with regard to the earmarking of funds must be introduced.

Clearly, priority should be given to soil and water conservation, waste-land development and social forestry followed by rural roads and rural housing. No doubt, inter-se importance will vary from place to place. The present system of earmarking a certain quantum for Million Wells Scheme and for housing under Indira Awaas Yojana would have to be relaxed, since several State Governments are not in a position to fulfil these stipulations. Furthermore, given the paucity of resources, one would perhaps have to concentrate the resources under the JRY to the more backward districts so as to reach the poorest.

This would require an assessment of the extent and nature of unemployment and underemployment and the local level requirements at the village level. Even though a large proportion of the JRY funds flow directly to the village, activities undertaken should be such as to fulfill local needs within the overall framework of the village plan.

Under the IRDP, assistance is given to individual beneficiaries for acquisition of an asset. While one-third is in the form of subsidy, two-thirds are in the form of bank loans. Hence, the banks need to assess the economic viability of an asset before giving assistance. However, the entire focus on targetting makes such an exercise futile.

Actually, the matter should be viewed not from the supply side but from the demand side i.e. identifying activities which are appropriate, given the skills of the beneficiaries, the infrastructure and the linkages available. Wherever necessary skills are not of the required standard, this upgradation should be facilitated under TRYSEM. In other words, IRDP needs to be viewed as a credit based self-employment programme with an element of subsidy rather than as a programme based on subsidy supplemented by bank credit. Under DWCRA, the results have not been quite satisfactory. While the idea of organising women into groups to take up activities which yield supplementary income is a sound one it has suffered on account of lack of adequate investment and selection of unviable activities.

Therefore, it may be worthwhile to encourage formation of thrift and credit societies which will be entitled to receive matching contributions from the

Government. This is already being attempted in some States. There is also an urgent need for conscientisation of rural women through activists, social workers and voluntary agencies. Women need to be encouraged to form cooperatives or institutions of self-employed women in order to become viable groups.

In addition, child care, reduction in drudgery and organisation of women beneficiaries need to be pursued more vigorously. Marketing of products made by women's groups is an important aspect and State Governments must make provisions for purchase of their products by various Government departments, emporia, and through melas and fairs.

Upgradation of skills and technology need to be given a special thrust with the aim of generating employment in new areas where demand is expanding. The target for TRYSEM trainees has been doubled from about 20 to 40 lakhs per annum. However, in order that those trained could find employment it is necessary that (a) training needs are assessed in terms of activities which can be either started under ERDP or in such fields where there is likely to be an increase of wage employment opportunities, (b) the quality of training should be such as to bring about improvement in the skill endowment of the trainees, (c) groups of persons can be organised in a particular trade or productive venture and these can be brought together for training.

## Integration of Poverty Alleviation Programmes

The programmes themselves need not be changed but the manner of implementing them would need some modification. A high degree of convergence can be attempted in a few districts on a pilot basis by an integration of the poverty alleviation programmes, the area development programmes and sectoral schemes.

Taking a district as a unit of planning, a district plan would need to be prepared, taking into account the physical and human endowments of that area, the felt needs of the people and the funds available. Projects and schemes would be selected for implementation based on these. Using scientific methods now available, the geographical area of the district would have to be mapped from photogrammetry and satellite data.

The maps would then be subjected to analysis for identification of water harvesting structures such as aalabands, gulley plugs, infiltration galleries and terraces. Such a strategy would ensure that soil erosion would be minimised and surface run off is virtually eliminated, with conservation of every drop of rainfall.

This district map would then have to be disaggregated at the village level. Viable activities particularly in agriculture and allied sectors including animal husbandry, pisciculture, horticulture forestry and agro-processing would have to be selected. Village and small industries with potential can also be identified for priority. In addition, development of infrastructural support and forward and backward linkages will have to be ensured which are essential prerequisites for the viability of the selected activities.

Emphasis on human resource development would have to be placed, as productivity depends both on natural resources and on the level of human resource development. Therefore, it will be necessary to integrate the social aspects of development including education, health and access to safe drinking water with the plan for economic development.

Briefly, therefore, the strategy would consist of creating the right environment for success of family plans rather than the present practice of farming them out and assuming that the infrastructure would look after itself. For planning and implementation of the district plan, the responsibility would vest in the Zilla Parisahds where they exist and/or in the DRDAs.

However, this should not preclude the involvement of people's representatives both elected and non-elected and voluntary organisations from taking on the planning and implementation of various schemes and programmes. In fact, it would be ideal to form committees involving representatives from both the Government and non-Governmental organisations. While surveys have been conducted in every village to identify those living below the poverty line, it is also important to have a registration of the unemployed and underemployed who are willing and able to work. This is essential in order to have an idea of the number of people to be covered under the various anti-poverty programmes.

The above are the broad parameters of the Integrated Rural Development approach the details of which would vary from place to place. It-is proposed in the Eighth Plan that one district per State is selected for the implementation of this programme in the manner described earlier.tions he set up at the central level to provide a forum for them.

In the Eighth Plan, a greater emphasis will be put on th role of voluntary organisations in rural development. A nation-wide network of NGOs will be created. In order to faciltate the working of this network, three schemes relating to She creation/replication/ multiplication and consultancy development have been worked out by the Planning Commission.Efforts will be made to evolve a system for providing one window service to NGOs wrking in the area of integrated development.

Land is still the single most important asset in rural India and given the present state of agricultural technology even a small farm can be viable, both in terms of employment and income of a family. The need for land reforms was recognised at the time of independence and has been reiterated in the successive Five Year Plans. The Seventh Plan enunciated land reforms to be an intrinsic part of the anti-poverty strategy.

The importance of land reforms continues to be significant. Its main tenets are abolition of intermediaries, security of tenure for tenant -cultivators, redistribution of land by imposition of a ceiling on agricultural holdings, consolidation ofholdingc and updating of land records. The Eighth Plan would therefore address itself to the factors that have come in the way of realising the goals of land reforms policy.

First, it would aim at ensuring that an atmosphere is created whereby the actual cultivators are made aware of their rights and enabled to claim their benefits. Secondly, it would encourage steps to be taken for early detection of surplus lands. Thirdly, it would be necessary to ensure that the newly acquired lands are brought under profitable agronomic practices, thus meeting the twin objectives of poverty alleviation and output growth. The management of land records and the skills and capabilities of the lower level official machinery would need to be given the necessary support of resources and modernisation so that they help, rather than hinder, the evolution of an equitable agrarian order.

On the question of tenancy, there are three aspects which need to be examined in some detail. First, the need to inculcate among tenants a degree of solidarity so that they can, at a time, counter the dominance of the landed classes as well as make the revenue administration accountable to themselves. Secondly, the mechanisms for transfer of title to the actual cultivator will require to be professional and sensitive. The third, and perhaps the most important aspect, is to make the gains real by getting from the land the quickest returns via access to a package of modem input.

Measures would be taken to make real the gains of tenancy laws by restricting the right to resumption; tackle absentee landlordism by defining personal cultivation more precisely and reviewing the provisions for regulating voluntary surrender. The National Commission on Revitalisation of Revenue Administration will take up all issues relating to land record management in the States. Organisations of tenants and sharecroppers will engine detection of informal and concealed tenancies, bring on record the tenants and share croppers.

On the question of land ceilings, the two aspects needing urgent attention ay; a) detection of surplus lands, hitherto unavailable because of recourse to evasive methods like benami transfers, partitions, fraud, collusion with official machinery etc., and b) ensuring that the allottees retain possession and there is severe penalty for dispossession. The lacunae in the laws will have to be removed so as to help resolve both these issues.

Suitable creative options need to be built into the law so that once the land is declared surplus, unless mala fide is established against the official machinery concerned, the land would vest in the Government, and it would be open to the Courts to award only compensation to the landlord. In respect of consolidation of holdings, two aspects that need attention are; a) The smaller farmers harbour strong apprehensions about getting a raw deal in the process of exchanging parcels of land towards the consolidation of holdings and b) The process of breaking up of holdings is a continuous one and a one-time settlement does not really solve the problem.

The solution, will therefore lie in making the farmer recognise that it is advantageous to share income from land rather than the land itself. The modality of bringing these aims into the land-related customs and practices will be more effective than the passing of laws. The common property resources have traditionally been a source of economic sustenance for the weaker sections of society.

Measures will be taken to survey their extent so that the encroachments by more influential sections can be removed. This is an area where voluntary organisations and local democratic institutions will be associated with the administrative machinery to restore to the panchayat/commu-nity the ownership of common property resources so that further encroachments do not take place.

Efforts will be made to develop these resources so that the option is once again open for the poorer sections to exploit for supplementing their income. In order to create a reward system for the better performing States in these matters, it is proposed to set up an index of performance with regard to land reforms. Based on this indicator it will be possible to provide a portion of the general pool of Central Assistance to the states.

# Bibliography


Abed, Fazle Hasan, and Imran Matin. 2007. Beyond Lending: How Microfinance Creates New Forms of Capital to Fight Poverty. *Innovations* 2 (1–2): 3–17.

Ali, Ifzal. 2007. Pro-Poor to Inclusive Growth: Asian Prescriptions. *ERD Policy Brief Series* No. 48. Manila: ADB.

Ban, Sung Hwan, et al. 1980. *Rural Development: Studies in Modernization of the ROK: 1945–1975*. Cambridge: Harvard University Press.

Bolt, Richard. 2004. Accelerating Agriculture and Rural Development for Inclusive Growth: Policy Implications for Developing Asia. ERD Policy Brief Series No. 29. Manila: ADB.

Campos, Jose Edgardo, and Hilton L. Root. 1996. *The Key to the Asian Miracle: Making Shared Growth Credible*. Washington DC: The Brookings Institution.

Chambers, Robert. 1983. *Rural Development: Putting the Last First*. London: Longmen.

Chaudhuri, S., and M. Ravallion. 2007. Partially Awakened Giants: Uneven Growth in China and India. In *Dancing with Giants: China, India and the Global Economy*, edited by L.A. Winters and S. Yusuf. Washington DC: World Bank.

Consultative Group to Assist the Poor (CGAP). 2006. *Graduating the Poorest Into Microfinance: Linking Safety Nets and Financial Services*. Focus Note No. 34. Washington DC: CGAP.

de Haan, Arjan, and Michael Lipton. 1998. Poverty in Emerging Asia: Progress, Setback, and Log-jams. *Asian Development Review* 16(2): 135–176.

Fan, Shenggen, et al. 2000. Government Spending, Growth and Poverty in Rural India. *American Journal of Agricultural Economics* 82(4): 1038–1051.

Felipe Jesus, and Rana Hasan, ed. 2006. *Labor Markets in Asia: Issues and Perspectives*. New York: Palgrave Macmillan.

Food and Agriculture Organization (FAO). 2006. *The State of Food andAgriculture.* Rome: FAO.

Geyer-Allly, Elaine. 1994. Agriculture, Technology and the Environment in OECD Member Countries. In OECD, *Agriculture and the Environment in the Transition to a Market Economy.* Paris: OECD.

Lipton, Michael. 1977. *Why Poor People Stay Poor: Urban Bias in World Development.* London: Temple Smith.

McKinon, Ronald I. 1973. *Money and Capital in Economic Development.* Washington DC: The Brookings Institution.

Pulley, Robert V. 1989. Making the Poor Creditworthy: A Case Study of the Integrated Rural Development Program in India. World Bank Discussion Papers. Washington DC: World Bank.

Rao, P. Parthasarathy, et al. 2006. Diversification towards High Value Agriculture: Role of Urbanization and Infrastructure. *Economic and Political Weekly* June 30: 2748–2753.

Ravallion, Martin. 1991. *Reaching the Rural Poor through Public Employment: Arguments, Evidence and Lessons from South Asia.* World Bank.

Rosegrant, Mark W., and Peter B. R. Hazell. 2000. *Transforming the Rural Asian Economy: The Unfinished Revolution.* Hong Kong: Oxford University Press.

Saith, Ashwani. 1987. Contrasting Experiences in Rural Industrialization: Are the East Asian Success Transferable? In *Rural Industrialization and Employment in Asia, edited by Rizwanul.* Islam New Delhi: International Labour Organization.

Sinha, Sanjay, et al. 2006. *Microfinance in South Asia: Towards Financial Inclusion for the Poor.* Washington DC: World Bank.

Thomas, Mark, et al. 2006. Rural Development in Asia. Paper prepared for the Asian Development Bank. London: Emerging Market Economies Ltd.

Thorat, Sukhadeo, and Shenggen Fan. 2007. Public Investment and Poverty Reduction: Lessons from China and India. *The Economic and Political Weekly* 42(8): 704–710.

United Nations Development Program (UNDP). 2005. *Human Development Report 2005.* New York: UNDP.

World Bank. 1978. *Rural Enterprises and Nonfarm Employment.* Washington DC: World Bank.

# Index